Cosmetic Products and Industry - New Advances and Applications

Edited by Usama Ahmad and Juber Akhtar

Published in London, United Kingdom

Cosmetic Products and Industry - New Advances and Applications
http://dx.doi.org/10.5772/intechopen.105296
Edited by Usama Ahmad and Juber Akhtar

Contributors
Johnbrynner García, Ricardo Alejandro Pineda Beltrán, Sara Gonçalves, Isabel Gaivão, Júlia Scherer Santos, Guilherme Diniz Tavares, Thaís Nogueira Barradas, Mirsiane Pascoal Costa, Carolina Sousa Ponciano, Irina Saretzky, Marta Cassini, Cristian Rosales, Usama Ahmad, Shivangi Sharma, Juber Akhtar, Anas Islam, Mohd Muazzam Khan, Naureen Rizvi

First published in London, United Kingdom, 2023 by IntechOpen
IntechOpen is the global imprint of INTECHOPEN LIMITED, registered in England and Wales, registration number: 11086078, 5 Princes Gate Court, London, SW7 2QJ, United Kingdom

British Library Cataloguing-in-Publication Data
A catalogue record for this book is available from the British Library

Additional hard and PDF copies can be obtained from orders@intechopen.com

Cosmetic Products and Industry - New Advances and Applications
Edited by Usama Ahmad and Juber Akhtar
p. cm.
Print ISBN 978-1-83768-622-3
Online ISBN 978-1-83768-623-0
eBook (PDF) ISBN 978-1-83768-624-7

For EU product safety concerns:
IN TECH d.o.o., Prolaz Marije Krucifikse Kozulić 3, 51000 Rijeka, Croatia,
info@intechopen.com or visit our website at intechopen.com.

Meet the editors

Dr. Usama Ahmad holds a specialization in Pharmaceutics from Amity University, Lucknow Campus, India. He received his Ph.D. from Integral University, India, in 2018. Currently, he is an Assistant Professor of Pharmaceutics in the Faculty of Pharmacy, Integral University. From 2013 to 2014, he worked on a research project funded by the Science and Engineering Research Board-Department of Science and Technology (SERB-DST), Government of India. He has a rich publication record with more than 30 original journal articles, 3 edited books, 4 book chapters, and several scientific articles published in *Ingredients South Asia Magazine* and *QualPharma Magazine*. He is a member of the American Association for Cancer Research, the International Association for the Study of Lung Cancer, and the British Society for Nanomedicine. Dr. Ahmad's research focus is on the development of nanoformulations to facilitate the delivery of drugs that aim to provide practical solutions to current healthcare problems.

Dr. Juber Akhtar obtained his BPharm from Jamia Hamdard University, New Delhi, India, in 2005. He obtained his MPharm from Manipal University, Karnataka, India, in 2007 and his Ph.D. from Integral University, India in 2014. He is currently a professor and head of Department of Pharmacy, Integral University. Dr. Akhtar also served as chairman cum biological scientist for IAEC from 2015 to 2017. He also has experience teaching abroad and has served as a professor at Buraydah College of Pharmacy and Dentistry, Kingdom of Saudi Arabia. Dr. Akhtar has more than 70 publications in reputed journals to his credit and is an editorial member of many esteemed journals. He has supervised a dozen Ph.D. and MPharm students in research projects. Dr. Akhtar is actively involved in research activities and his areas of interest include the development of nano-particulate drug delivery systems for targeting various organs.

Contents

Preface **IX**

Chapter 1 **1**
The Art and Science of Cosmetics: Understanding the Ingredients
by Shivangi Sharma, Usama Ahmad, Juber Akhtar, Anas Islam, Mohd Muazzam Khan and Naureen Rizvi

Chapter 2 **25**
Skin Care Nanocosmetics
by Júlia Scherer Santos, Carolina Sousa Ponciano, Thaís Nogueira Barradas, Mirsiane Pascoal Costa and Guilherme Diniz Tavares

Chapter 3 **37**
The Beauty and the Toxic Beast: Use of Comet Assay to Study Antigenotoxicity of Natural Ingredients
by Sara Gonçalves and Isabel Gaivão

Chapter 4 **53**
Enhancing Skin Cicatrization with Natural Sources – The Role of Polyunsaturated Fatty Acids (PUFAs) and Beeswax
by Irina Saretzky and Marta Cassini

Chapter 5 **81**
Thermal Behavior of Waxes and Its Correlation with Mascara Stability Tests: A DSC Study
by Ricardo Alejandro Pineda Beltrán and Johnbrynner García

Chapter 6 **91**
A Novel Method for Titanium Dioxide Quantification in Cosmetic Products via Borate Fusion by Flame Atomic Absorption Spectroscopy
by Cristian Rosales and Johnbrynner García

Preface

In the ever-evolving landscape of beauty and personal care, the cosmetic products industry stands as a testament to human creativity, innovation, and the pursuit of enhancing both our appearance and self-confidence. The journey of cosmetics, from ancient rituals to modern science, has been a remarkable exploration of nature's resources and our ability to harness them for personal adornment. The purpose of this book, *Cosmetic Products and Industry – New Advances and Applications*, is to illuminate the multifaceted world of cosmetics, shedding light on its technological advancements and the exciting possibilities that lie ahead.

This book provides a panoramic view of this dynamic landscape. The contributing authors, experts in their respective fields, have joined forces to present a collection of chapters that offer both depth and breadth of insight. Whether you are a seasoned industry professional, an aspiring cosmetic scientist, or simply someone curious about the science behind the products you use every day, this book is a valuable resource.

Chapter 1 "The Art and Science of Cosmetics: Understanding the Ingredients" by Dr. Usama Ahmad et al. provides a comprehensive overview of the use of cosmetic ingredients, their specific effects on product development and properties, emerging alternatives in the personal care market, and the advancement of personal care products. The authors discuss in detail the key ingredients used to formulate stable and effective cosmetic products.

Chapter 2, "Skin Care Nanocosmetics" by Júlia Scherer Santos et al., reviews several nanotechnology-based cosmetics that have been developed to achieve more effective products, with a focus on vesicular carriers. The authors highlight that beyond the benefits provided by nanocarriers, nanotechnology-based cosmetics have attracted the cosmetic industry's interest because they have high market value, which may reflect an increased market consumption. Skin delivery of cosmetic ingredients via nanocarriers provides a better alternative to traditional cosmetics by improving skin hydration, skin penetration, stability, and versatility.

Chapter 3, "The Beauty and the Toxic Beast: Use of Comet Assay to Study Antigenotoxicity of Natural Ingredients" by Sara Gonçalves and Isabel Gaivão, presents an important research study to demonstrate, in vivo, the genotoxicological effects of elderberries, almonds, olives, and grapes in the Drosophila melanogaster species. The results show that all four natural ingredients had antigenotoxic effects, meaning that they protected fruit fly DNA from damage. Elderberry was the most potent antigenotoxic agent, followed by almond, olive, and grape. These findings suggest that natural ingredients may be used to develop new cosmetics with potential health benefits.

Chapter 4, "Enhancing Skin Cicatrization with Natural Sources – The Role of Polyunsaturated Fatty Acids (PUFAs) and Beeswax" by Irina Saretzky and Marta Cassini, presents a study in which the authors reviewed the skin cicatrization process, chronic wounds, and clinical cases treated with a combined formula rich in PUFAs and beeswax. The formula was developed by the authors of the study and is based on the synergistic effects of PUFAs and beeswax. The study found that the formula was effective in the treatment of chronic wounds, including diabetic foot ulcers, pressure ulcers, and venous ulcers. The formula was also safe and well tolerated by patients. The findings of this study suggest that natural products may be a safe and effective alternative to conventional wound care treatments. However, more research is needed to confirm these findings and to optimize the use of natural products in wound care.

Chapter 5, "Thermal Behavior of Waxes and Its Correlation with Mascara Stability Tests: A DSC Study" by Ricardo-Alejandro Pineda-Beltrán and Johnbrynner García, presents valuable research work on the effect of heat-sensitive waxes on the stability of mascara formulas. The researchers found that three of the eight commercial mascaras studied showed visible separations. The authors suggest that avoiding heat-sensitive waxes in mascara formulas can prevent visual phase separations and produce thermally stable mascaras. The findings of this study are important for mascara manufacturers because they can help to improve the stability of mascara formulas and prevent visual phase separations, which can lead to a better user experience and a more consistent product.

Chapter 6, "A Novel Method for Titanium Dioxide Quantification in Cosmetic Products via Borate Fusion by Flame Atomic Absorption Spectroscopy" by Cristian Rosales and Johnbrynner García, presents research on the development of a new method to extract titanium dioxide (TiO_2) from cosmetic products. The method uses borate salts and flame atomic absorption spectroscopy (FAAS) to quantify the amount of TiO_2 in the product. The developed method was tested on a variety of cosmetic products, including oil-in-water and water-in-oil emulsions, as well as foundations and lipsticks. The results show that the method is specific, linear, sensitive, precise, and accurate. The method is also safer and more easily transferable than other methods for analyzing TiO_2 in cosmetics. This makes it a convenient tool for routine analysis of cosmetic products.

This book is a valuable resource for researchers and professionals in the cosmetic industry. It provides a comprehensive overview of the latest research on cosmetic ingredients and their applications. The book is also a valuable reference for students and anyone interested in learning more about the science of cosmetics.

Usama Ahmad and Juber Akhtar
Department of Pharmaceutics,
Faculty of Pharmacy,
Integral University,
Lucknow, India

Chapter 1

The Art and Science of Cosmetics: Understanding the Ingredients

Shivangi Sharma, Usama Ahmad, Juber Akhtar, Anas Islam, Mohd Muazzam Khan and Naureen Rizvi

Abstract

Cosmetics, a captivating category of over-the-counter products that enhance appearance and promote skin health, have become ubiquitous in modern society. Initially associated with altering one's appearance, cosmetics have evolved beyond their traditional beauty role to encompass skincare and the treatment of various skin conditions. While numerous chemicals can be utilised in cosmetic formulations, key ingredients include water, emollients, humectants, surfactants, preservatives, antioxidants and ultraviolet (UV) filters. With the rise in consumer preferences for clean beauty, silicone- and paraben-free cosmetics and cruelty-free testing, safe and effective herbal and plant-based products have gained significance. Researchers are actively exploring the development of natural cosmetics, leveraging the synergistic properties of these substances. Nanoformulations such as liposomes, nanoparticles and emulsions have been investigated and proven safer and more efficient than conventional cosmetics. This chapter provides a comprehensive overview of the utilisation of these ingredients in cosmetics, their specific effects on product development and properties, emerging alternatives in the personal care market, the widespread adoption of nanotechnology in various scientific fields, and its impact on the advancement of personal care products.

Keywords: cosmetics, personal care product, ingredients, safety, nanotechnology

1. Introduction

Cosmetics today have become an essential, unavoidable necessity to an increasing number of individuals; they are now a part of the daily routine of every next individual, and huge amounts are being consumed every year [1]. Considering and examining the history of cosmetics, it is believed that cosmetics began in the Stone Age, mainly by Egyptians [2]. When considering the utilisation of cosmetics in the contemporary era, the primary and evident purpose is to safeguard the body against various natural elements, such as sunlight, rain, insects and more. One of the primary reasons for the widespread use of cosmetics in modern society is to ensure personal hygiene. Cosmetics serve the purpose of enhancing and promoting attractiveness through makeup, as well as providing skin and hair care. This is particularly important in protecting against various pollutants, sunlight and other environmental factors. These factors have the potential to cause damage or deterioration to the skin and hair [3].

Hence, cosmetology is intended to be applied to the human body or any parts for beautifying, cleansing, making it look attractive, promoting, or modifying its appearance [4]. Numerous self-care products are available in the market for different purposes, including lotion, creams, moisturisers, shampoos, soap, oils, etc. This chapter will cover the basics of cosmetics, how components impact the formulation of a cosmetic, the many ingredients that make up a cosmetic formulation, and the selection criteria used to make these choices. This chapter will provide comprehensive information about cosmetics, their ingredients and the part they play in creating an ideal cosmetic.

2. Role of ingreidents in cosemetic formulation

There are numerous approaches for discussing the substances used in cosmetics, but it makes the most sense to concentrate on their main function in the formulation. Any ingredient that is used to build a cosmetic product should be safe, non-irritant, non-toxic and effective [5]. Cosmetic products incorporate various types of polymers in their formulations in accordance with the tasks for which they are designed since polymers are essential to the creation of cosmetics [6, 7].

Numerous cosmetic chemicals have a skincare effect. These compounds have a variety of chemical attributes that function through their own unique principles to care for our skin, making it more moisturised, firm, smooth and radiant, among other things. Nowadays, there's a noticeable trend of individuals transitioning to cosmetics incorporating plant-based ingredients. Moreover, people are increasingly seeking cruelty-free formulations devoid of harmful chemicals, silicones and parabens. The more greener and clean the cosmetic product is, the more it is accepted by the consumers like in place of surfactant. Biosurfactants can be used in the pharmaceutical business as a green substitute. They can make medications more soluble, especially ones that are not water-soluble [8]. A class of surface-active chemicals known as "biosurfactants" is generated from microorganisms like bacteria, fungi and yeast [9]. They have attracted a lot of interest since they are superior to synthetic surfactants in that they are made from renewable resources, have low or non-toxic side effects, are biodegradable, have good surface activity, are highly specific, and work well in circumstances of extreme heat and pH [10].

2.1 Types of ingredients in cosmetics

For formulating a cosmetic formulation, several ingredients are broadly divided into two categories: functional ingredients and performance ingredients.

2.1.1 Functional ingredients

These are the ingredients that are necessary for the product to function as intended, such as emulsifiers, thickeners and preservatives. One of the major responsibilities of the emulsifier is to assist in creating the interface between the continuous oil phase and water droplets. In a 2013 study, three water-in-oil body lotions were created using different emulsifier mixtures. Rheological investigations predict several sensory characteristics, including thickness and bottle-pouring behaviour [11]. A study in 2016 evaluated, after 4 weeks of use on mature human skin, the effects of cosmetic formulations, eye creams and facial creams containing palmitoyl peptides, *Silybum marianum* seed oil, vitamin E and other functional ingredients on improvements in facial wrinkles, elasticity, dermal density and skin tone. The outcomes demonstrated that the facial and eye cream's constituents, including palmitoyl peptides, *S. marianum* seed oil, vitamin E and others, impact the reduction of face wrinkles, elasticity, dermal density and skin tone [12].

2.1.2 Performance ingredients

These ingredients provide specific benefits to the skin or hair, such as moisturising, anti-ageing or UV protection. Performance ingredients such as moisturisers include humectants, occlusive and emollients. These agents are responsible for the main properties of any given cosmetics. Occlusive and humectant are combined in moisturisers to increase the skin's ability to retain moisture. In a study conducted in 2020 titled "A Nature-Based Bakuchiol Anti-Ageing Moisturiser for Sensitive Skin, "a monoterpene from the seeds of *Psoraleacorylifolia* called bakuchiol functions like retinol in the regulation of gene expression. The outcome demonstrated that people with sensitive skin may tolerate and benefit from bakuchiol, a natural anti-ageing moisturiser [13]. Similarly, UV protection ingredients like titanium oxide and avobenzone are added to cosmetic formulations like sunscreen to shield the skin from damaging UV rays. An in vitro evaluation of sunscreen's broad-spectrum UV protection was conducted in a study in 2000; the method eliminates the need to expose volunteers to acute exposures of high-dose, non-terrestrial UV, the risks of which to human health are still poorly understood. It also provides a routine yet sensitive method of differentiating and classifying sunscreen products [14].

2.1.3 Natural ingredients

These are ingredients derived from natural sources, such as plants, minerals or animals. They are often perceived as safer and more environmentally friendly than synthetic ingredients. However, they can also be less stable and have a higher risk of contamination or allergic reactions. There are equal studies proving these natural ingredients' efficacy, safety and risk. A study was conducted in 2020 in which lignins from hazelnut and walnut shells were assessed as prospective bioactive components for cosmetic products. These lignins were investigated as potential natural active ingredients for healthcare products. The study yielded mixed results, with isolated lignins exhibiting UV-absorbing properties but the sun protection factor provided by lignin incorporation in pure cream falling short of the current standards established for the prevention of skin damage caused by sun exposure. The findings indicated the need for a strategy to improve the lignin polymer's UV performance [15]. A study done in 2017 investigated the amount of waste produced by food industrial operations have significantly increased, creating sustainability issues. This is due to the potential uses of bioactive compounds from grape processing byproducts as active ingredients for skin care products. These environment-friendly raw ingredients could help the so-called "eco cosmetic." The evaluation found that for a winery to be involved in the cosmetic sector, a reliable supply of materials must be provided, keeping in mind that the equipment and sustainable practices required for cosmetic incorporation should not be overlooked [16].

2.1.4 Synthetic ingredients

These are chemically synthesised ingredients in a laboratory. They are often more stable, consistent and cost-effective than natural ingredients. However, they can also be perceived as less safe and less environmentally friendly than natural ingredients. Trans

Epidermal Water Loss (TEWL) was examined before, during, and after 15 and 30 days of application in a study in 2011 that evaluated the skin moisturising effectiveness of formulations containing various concentrations of panthenol. It was determined that the formulation's improved protective effect of 1.0% panthenol added to the formulation maintains skin integrity. The main reasons synthetic compounds are used in formulations are their quick action, economical nature, and easy accessibility [17].

3. Common cosmetic ingredients and their function

Countless cosmetic products are available, each with a unique combination of components. A typical product will contain anything from 15 to 50 ingredients as shown in **Figure 1**.

3.1 Water

Since water is a universal solvent, it can be combined with emulsifiers and "thicker" components like butter and oils to create emulsions that are used to make creams and lotions. A study in 2018 created a cosmetic formulation of acai berries using an oil-in-water emulsion using ultrapure water [18]. Similarly, a study in 2018 employed rice water as the primary ingredient to construct a semisolid dosage form with cosmetic qualities for topical application and found it beneficial against skin ageing. Recently, the search for new bioactive substances to prevent skin ageing has grown [19].

3.2 Emollients

Emollients are a class of ingredients used in cosmetics formulations to soften, moisturise and condition the skin or hair. They work by forming a protective barrier

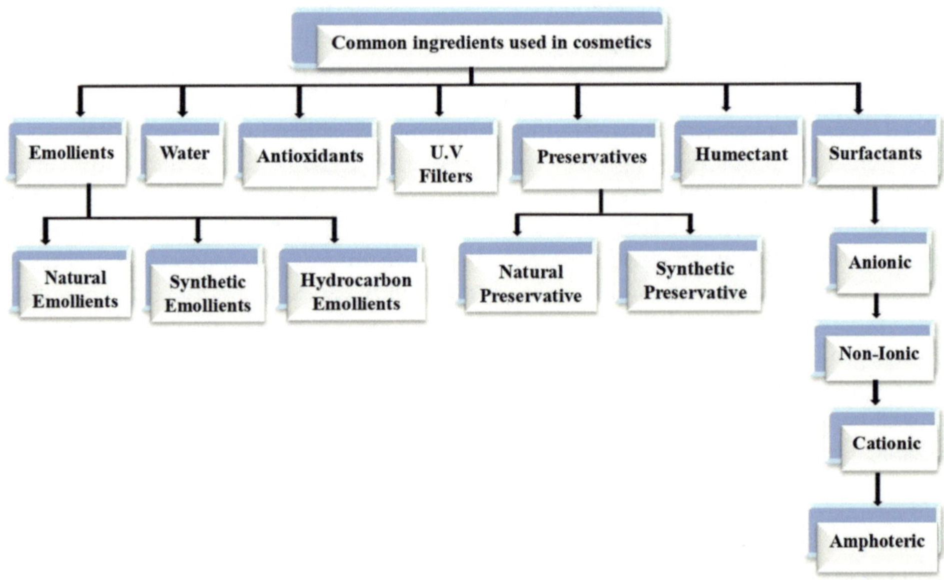

Figure 1.
Common ingredients used in cosmetics.

on the skin or hair, reducing moisture loss and preventing dryness and cracking. Emollients can be derived from natural or synthetic sources and come in various forms, such as oils, butter and waxes.

3.2.1 Classification of emollients

Emollients can be classified based on their source of origin, chemical structure and physical properties. Here are some common classifications of emollients:

3.2.1.1 Natural emollients

Emollients in nature come from either plant or animal sources. They are good for nourishing and moisturising the skin or hair because they are frequently packed with vitamins, antioxidants and fatty acids. Some examples of natural emollients include Shea butter, coconut oil, jojoba oil, etc. In a study published in 2017, authors prepared poly herbal creams using almond oil and olive oil for their moisturising properties [20]. In another study in 2020, they prepared herbal cosmetic cream of *Urticadioica, Amaranthus viridis* and *Aloe vera* using shea butter, coconut oil, germanium oil, jasmine oil, tea tree oil and honey as emollients, and they found the formulation was potent as an antibacterial agent, safe for application on the skin and cosmetically appealing to enhance patient/customer compliance [21]. Another research done in 2020 formulated lip balm using these natural emollients: almond oil, virgin coconut oil and honey and the lip balms produced based on the natural ingredients were found to be safe to use [22].

3.2.1.2 Synthetic emollients

Synthetic emollients are created in a laboratory and are often used in cosmetics formulations because they are cost-effective and have specific properties. Some examples of synthetic emollients include Dimethicone, Isopropyl myristate and Glyceryl stearate. In a study in 2014, they formulated oil-in-water emulsions with cosmetic properties using glyceryl stearate. The results showed Glyceryl stearate to be the dominant co-emulsifier affecting emulsion properties [23]. In a study in 2015, researchers formulated cosmetic multiple emulsion (W/0/W) using cetyldimethiconecopolyol; the result indicated that 2.4% cetyldimethiconecopolypol stabilises the emulsion [24]. A study in 2021 designed a squalane and dimethicone-based formula for treating head lice. The results indicated that the formula was safe for children with good skin compatibility and acceptability [25].

3.2.1.3 Hydrocarbon emollients

Hydrocarbon emollients are derived from petroleum and are commonly used in cosmetics formulations because of their low cost and excellent skin-softening properties. Some examples of hydrocarbon emollients include mineral oil and petrolatum. Both of these has been used in the development of various cosmetic formulation; in a study in 1996, white mineral oil was shown to be a highly effective material for adding moisturisation and other beneficial characteristics to the skin when used [26]. Similarly, in 2001, a research work investigated how well a petrolatum-based o/w emulsion penetrated the skin, and the results revealed that the level of penetration of the sunscreen ingredients by the epidermis depends on the type of vehicle the sunscreen agents are contained in [27].

Emollients can be classified based on their chemical structure into the following categories.

3.2.1.4 Fatty acids and derivatives

This category includes emollients such as stearic acid, palmitic acid and their derivatives, such as stearates and palmitates. These emollients are derived from natural sources such as animal and vegetable fats and oils. In some studies, like one in 2010, they formulated a mild cleanser using stearic acid, which showed excellent deep cleansing properties [28].

3.2.1.5 Fatty alcohols

Examples of this category include cetyl alcohol, stearyl alcohol and behenyl alcohol. These emollients are also derived from natural sources and are typically used to thicken and stabilise formulations. Cetyl alcohol was utilised in a study in 2008 to formulate sunscreen, and the findings of this study demonstrated that sunscreen in vitro skin penetration was significantly influenced by the type of emulsifying system employed to create O/W emulsions [29].

3.2.1.6 Esters

This category includes a wide range of emollients, including isopropyl myristate, isopropyl palmitate and octylpalmitate. Esters are synthetic compounds derived from the reaction between an acid and an alcohol. They are known for their light, non-greasy texture and their ability to penetrate the skin easily. In a study in 2006, they formulated a gel-cream containing *retinyl palmitate* and *tocopheryl acetate* and found that using these two increased the retention time and stability of the formulation [30].

3.2.1.7 Silicones

Examples of this category include dimethicone, cyclomethicone and phenyl trimethicone. These emollients are synthetic compounds that have a unique, non-greasy texture and offer a high degree of skin smoothing and softening. Silicones are widely used nowadays in every cosmetic formulation. Research is also being carried out to improve the use of silicones like a research work conducted in 2018 developed a hair care formulation containing silicones and vegetable oil, and the results showed that silicones helped in the treatment of hair fibres and made hair softer [31].

3.3 Humectants

These are ingredients that attract and retain moisture in the skin, such as glycerin, hyaluronic acid and urea. These ingredients attract water to the skin and help to keep it there. They work by drawing moisture from the environment and retaining it in the skin. Some common examples of humectants include hyaluronic acid, glycerine and urea. In a research work carried out in 2004, the investigation found that the newest hyaluronic acids were more long-lasting and more suitable for cosmetic treatments [32]. Similar to this, many other researches were carried out. According to a study in 2018, the preparation of nanoparticles quaternised cyclodextrin-grafted chitosan associated with hyaluronic acid for cosmetic use, the effectiveness of the chosen nanoparticles at moisturising was enhanced as a result of the formation of cross-linked networks in the nanoparticles [33]. For use in hair care cosmetics, investigators developed films based on a combination of collagen, chitosan and hyaluronic acids for use in hair care cosmetics. Combining chitosan, collagen, and hyaluronic acid forms a three-component mixture aimed at providing hair nourishment, potentially leading to increased hair thickness and improved mechanical properties [34]. Similarly, a study published in 2012 used glycerine to formulate an aloe vera gel cosmetic hydrogel; it was observed that glycerine improved the viscosity, and the gel was found to be commercially used for cleaning, softening and improving the texture of the skin [35].

3.4 Surfactants

Surfactants, also known as surface-active agents, are compounds that lower the surface tension between two immiscible substances, such as water and oil. They have a hydrophilic (water-loving) and a lipophilic (oil-loving) end, which allows them to interact with both polar and non-polar substances [36].

3.4.1 Types of surfactants

3.4.1.1 Anionic surfactants

Anionic surfactants are the most common type of surfactant and have a negatively charged head group. Anionic surfactants are widely used in cleansing products, such as shampoos, body washes and facial cleansers. They are effective at removing dirt, oil and other impurities from the skin and hair.

Examples of anionic surfactants in cosmetics include sodium lauryl sulfate (SLS) and sodium laureth sulfate (SLES) in shampoos and body washes, sodium dodecylbenzene sulfonate (SDBS) in facial cleansers and sodium lauryl sarcosinate (SLS) in toothpaste and mouthwash. Numerous studies are still being conducted to examine the effects of these anionic surfactants. For example, a study published in 2021 examined the effects of sodium lauryl sulfate applied as a patch to human skin to examine its physiology and discovered that using mild surfactants with little impact on the physiology, skin integrity, and bacterial community microbiome could be useful for preserving skin integrity and microbial balance for a beneficial outcome [37]. Another study published in 2019 compared the effect of anionic surfactant sodium lauryl sulfate vs. Sodium Coco Sulfate with respect to their Interaction with the skin and the result showed both the surfactant affected the metabolism of human keratinocytes [38].

3.4.1.2 Cationic surfactants

Positively charged cationic surfactants are frequently found in hair care and fabric softener products. They are frequently found in shampoos, conditioners and hair styling treatments for use on the hair. They work well to condition hair and get rid of frizz.

Examples of cationic surfactants in cosmetics include cetrimonium chloride in hair conditioners and detanglers, cocamidopropylbetaine (CAPB) in shampoos and body washes and behentrimoniummethosulfate (BTMS) in hair conditioners and leave-in treatments. The cosmetic ingredient review (CIR) expert panel carried out a study on the Safety Assessment of cocamidopropylbetaine (CAPB) and found it to be safe in cosmetics as long as they are formulated to be non-sensitising [39]. The results of a study conducted in 2008 on the sensitisation rate of cocamidopropylbetaine by patch testing in patients at a university hospital in Beijing revealed that the substance caused aesthetic dermatitis [40]. In a study in 2008, improved sustainability and highly concentrated cocamidopropylbetaine for Enhanced Skin Care discovered that more quality advancements and even extra skin care results, such as improved skin moisturisation, could be made [41]. A study published in 1998 conducted safety research on cocamidopropylbetaine and discovered that although this mild surfactant can be utilised in cosmetic products, it is advised that cocamidopropylbetaine with low levels of contamination be used when creating personal care products [42].

3.4.1.3 Nonionic surfactants

Detergents, shampoos and lotions all use nonionic surfactants, which have no charge on their head group and are frequently found in a variety of goods. They are adaptable and work well in both alkaline and acidic settings. They are effective at solubilising oils and other hydrophobic ingredients.

Examples of nonionic surfactants in cosmetics include Polysorbate 20 and Polysorbate 80 in facial cleansers and makeup removers, Decyl glucoside in baby washes and shampoos and polyethylene glycol (PEG) in moisturisers and sunscreens. Studies regarding the safety of various nonionic surfactants were carried out like in a study in 2015, which found that the polyethylene glycol, their ethers, and their fatty acid esters produce little to no ocular or dermal irritation and have extremely low acute and chronic toxicities [43]. A study in 2019 did a patch test study of Decyl glucoside and found they came up with the result that includes alkyl glucoside in all cosmetic series [44].

3.4.1.4 Amphoteric surfactants

Amphoteric surfactants have both positive and negative charges on their head group and are frequently included in gentle personal care products including body washes, face cleansers and infant shampoos. They are less likely to irritate the skin or eyes than other types of surfactants.

Here are some examples of amphoteric surfactants commonly used in cosmetics: Cocamidopropylhydroxysultaine (CHSB) is found in baby shampoos and body washes, Disodium cocoamphodiacetate (DSCA) is used in facial cleansers and body washes, Betaine is present in hair care products and facial cleansers, and Amphosol CA is utilised in body washes and facial cleansers. In a study published in 2020, investigators developed a foaming shampoo base for the treatment of seborrheic dermatitis and found that disodium cocoamphodiacetate was a good foaming surfactant in the amphoteric category [45].

3.5 Preservatives

Preservatives are substances added to cosmetics to prevent the growth of harmful microorganisms such as bacteria, fungi and viruses.

3.5.1 Importance of preservatives in cosmetics

a. Prevent microbial growth: Preservatives prevent the growth of harmful microorganisms such as bacteria, fungi and viruses, which can cause spoilage, discoloration and odour in cosmetic products.

b. Extend shelf life: Preservatives help extend the shelf life of cosmetic products, ensuring that they remain stable and safe for use over an extended period.

c. Maintain product quality: Preservatives help maintain the quality of cosmetic products, preventing changes in texture, colour and scent due to microbial growth or other factors.

d. Ensure regulatory compliance: Preservatives enable cosmetic manufacturers to produce high-quality, consistent products that meet regulatory requirements for microbiological safety and stability.

e. Prevent product recalls: The use of preservatives in cosmetic products reduces the risk of microbial contamination, which can lead to product recalls and damage to a company's reputation [46].

3.5.2 Types of preservatives

3.5.2.1 Natural preservatives

Plant-based Preservatives: Plant-based preservatives are derived from natural sources such as plants, fruits and vegetables. These preservatives are effective against a broad spectrum of microorganisms and are generally considered safe and non-toxic. Plant-based preservatives, such as rosemary extract, tea tree oil, and grapefruit seed extract, are commonly used in formulations. A study highlighted the diverse applications of grape extract and suggested it as an efficient, profitable, and eco-friendly alternative to other preservatives, with a focus on plant-based derivatives [47]. When combined with lavender and tea tree oils at a concentration of 0.5% each, a study in 2009 discovered that the antibacterial activity of tea tree oil and lavender oil may be increased while maintaining the same efficacy by reducing the preservative concentration by 0.1% [48].

Essential Oils: Essential oils are volatile compounds extracted from plants that have a characteristic aroma and are used in cosmetics for their fragrance and antimicrobial properties. Examples of essential oils used as preservatives in cosmetics include Lavender oil, clove oil, etc. A study published in 2013 found that essential oil has been found to show activity against a wide range of gram-positive and gram-negative bacteria [49].

3.5.2.2 Synthetic preservatives

Synthetic preservatives are chemical compounds synthesised in a laboratory and used in cosmetics to prevent the growth of microorganisms. Synthetic preservatives are generally more stable and have broader antimicrobial activity than natural preservatives. However, concerns have been raised about the safety of some synthetic preservatives, and some have been banned or restricted in cosmetic products. Here are some examples of commonly used synthetic preservatives in cosmetics.

Parabens: The most often used preservatives in cosmetics are parabens, which are p-hydroxybenzoic acid esters. They are frequently found in personal care products including shampoos, lotions and creams because they work well against a variety of pathogens. Methylparaben, propylparaben and butylparaben are a few types of parabens that are utilised in cosmetics. In vitro skin permeation and retention of parabens from cosmetic formulations were studied in a study published in 2007. They discovered that only the type of paraben, specifically its water solubility, affects skin penetration. The results also demonstrated that parabens can penetrate and accumulate in the skin [50].

Formaldehyde-releasing agents: Formaldehyde-releasing agents are preservatives that release small amounts of formaldehyde over time, which is effective against a broad range of microorganisms. Formaldehyde-releasing agents are commonly used in personal care products such as shampoos, conditioners and body washes. Examples of formaldehyde-releasing agents used in cosmetics include diazolidinyl urea, imidazolidinyl urea and quaternium-15. Many studies are carried out to check the safety of these formaldehyde-releasing agents; the risk of cosmetic formulations with formaldehyde above 0.2% is not negligible, according to a study in 2013 that used 1H NMR spectroscopy to measure the formaldehyde concentration in hair straightening products. These products may facilitate significant exposure to formaldehyde for consumers, especially for salon workers [51]. According to a study conducted in 2015 on the investigation into formaldehyde preservative release in cosmetic formulation, all preservatives in cosmetic matrices released significantly less formaldehyde than they did in pure aqueous or organic matrices. The formaldehyde-releasing amounts were also cosmetic-specific [52].

While formaldehyde-releasing agents are effective preservatives, they are also known as skin sensitisers and can cause allergic reactions in some people. As a result, some regulatory agencies have restricted the use of formaldehyde-releasing agents in cosmetic products.

Quaternary ammonium compounds: Quaternary ammonium compounds are cationic surfactants that have antimicrobial properties and are effective against a wide range of microorganisms, including bacteria, fungi and viruses. They are commonly used in personal care products such as shampoos, conditioners and body washes. Examples of quaternary ammonium compounds used in cosmetics include benzalkonium chloride, cetrimonium chloride and cetylpyridinium chloride. A study in 2020 evaluated benzalkonium chloride antimicrobial activity in a hand sanitiser, and the result showed that benzalkonium chloride reduced *Staphylococcus aureus* contamination [53]. While quaternary ammonium compounds are effective preservatives, they can also be skin irritants and can cause contact dermatitis in some people. As a result, some regulatory agencies have restricted the use of certain quaternary ammonium compounds in cosmetic products.

Organic Acids: Organic acids are weak acids frequently used in cosmetics as preservatives. They work well against a variety of microbes, including fungi and bacteria. Benzoic acid, sorbic acid and salicylic acid are a few examples of organic acids found in cosmetics. Organic acids are used in numerous cosmetic and hair care products like a study in 1998 created a salicylic acid-based peel for the treatment of photoaging; most patients notice smoother skin, less hyperpigmentation, and a reduction in fine wrinkles after the peel [54]. In a similar manner, investigators developed a salicylic acid peel and discovered for a variety of dermatological and aesthetic issues, such as acne vulgaris, melasma, photodamage, freckles and lentigines, salicylic acid is a safe and effective peeling agent. Dark skin types can utilise it without any problems [55]. While organic acids are generally considered safe for use in cosmetics, some may be skin irritants and can cause allergic reactions in some people.

Other synthetic preservatives used in cosmetics include phenoxyethanol, triclosan and chlorhexidine. These preservatives are effective against a broad range of microorganisms and are commonly used in personal care products.

Phenoxyethanol is glycol ether that is commonly used as a preservative in cosmetics. It is effective against bacteria and fungi and has low toxicity. However, concerns have been raised about its potential to cause skin irritation and contact dermatitis in some people.

Triclosan is a chlorinated aromatic compound that is effective against a wide range of microorganisms. It is commonly used in personal care products such as soaps, toothpaste and deodorants. However, concerns have been raised about its potential link to the development of antibiotic-resistant bacteria and its potential to disrupt endocrine function.

Chlorhexidine is a cationic bisbiguanide that is effective against a broad range of microorganisms, including bacteria and fungi. It is commonly used in oral care products such as mouthwash, toothpaste and some skin care products. However, concerns have been raised about its potential to cause skin irritation and allergic reactions in some people. In a study published in 2018, authors developed a new formulation of either chlorhexidine-containing toothpaste or mouthwashes for periodontal disease in order to study its clinical impact and cosmetic acceptability. The results revealed that subjects with periodontal disease who received oral care with a new formulation of either chlorhexidine-containing toothpaste or mouthwashes for 21 days reported a significant improvement in their symptoms and resolution of the gingivitis with no associated tooth discoloration [56].

3.6 Antioxidants

Vitamin E, vitamin C, and green tea extract are some examples of substances that shield the skin from oxidative damage produced by free radicals. The capacity of vitamin A and its derivatives to normalise keratinisation has been the main advantage of these ingredients in cosmetic goods. Vitamin A alcohol (retinol), vitamin A esters (retinyl palmitate, retinyl acetate), vitamin A aldehyde (retinal), and tretinoin (retinoic acid) are some of the common vitamin A compounds that can be found in cosmetics. These are present in cosmetic compositions in a variety of concentrations [57]. The water-soluble vitamin C, ascorbate, which is found in citrus fruits and vegetables, is significant for its antioxidant properties as well as its role as a cofactor in the hydroxylation processes that result in the formation of collagen. The capacity of vitamin C to directly squelch UV-induced free radicals and replenish vitamin E, another effective antioxidant, is one factor driving interest in vitamin C as a cosmetic element [58, 59]. Investigators formulated cosmetic multiple emulsion containing vitamin C and wheat extract and found that both of the creams behaved similarly from the dermatological point of view except for skin hydration as both the creams increased the hydration in the skin, the skin was shinier, and no sebum production occurred [60]. Solid lipid nanoparticles were created in a study published in 1999 as a cosmetic carrier for the use of vitamin E in cutaneous applications. The findings indicated that solid lipid nanoparticles are appropriate for integration in cutaneous formulations due to their great physical stability [61].

3.7 UV filters

These are ingredients that protect the skin from the harmful effects of UV radiation, such as zinc oxide, titanium dioxide and avobenzone. A sunscreen cosmetic

S.no	Category	Ingredients	Use/Comment	Reference
1	Emollients			[21, 22]
	Emollients based on source			
	Natural Emollient	Shea butter	Skin moisturisation boosting	
		Coconut oil	Exhibit Antibacterial properties	
		Jojoba oil	Greater moisturisation	
	Synthetic emollient	Dimethicone	Great co-emulsifier.	[24, 25]
		Glyceryl stearate	Emulsion stabiliser	
	Hydrocarbon emollient	Mineral oil	Greater moisturisation	[26, 27]
		Petrolatum	It was found to have excellent permeation in the epidermis	
	Emollients based on chemical structure			[28]
	Fatty acid & Derivatives	Stearic acid	Found to have excellent cleansing properties	
		Palmitic acid		
	Fatty alcohols	Cetyl alcohol	Good stabilisers Effect the formulation permeation rate	[29]
		Stearyl alcohol		
	Esters	Retinylpalmitate	They were found to increase the retention time and stability of the formulation	[30]
		Tocopheryl acetate		
	Silicones	Dimethicone	They have skin smoothing and softening properties were found to improve hair fibres and make hair softer	[31]
		Cyclomethicone		
2.	Humectants	Glycerin	Skin-softening properties	[33, 35]
		Hyaluronic acid	In a study, Hyaluronic acid nanoparticles were found to have improved moisturising efficacy	
3.	Surfactants			[37]
	Anionic Surfacant	Sodium lauryl sulfate (SLS)	Low amounts help to maintain skin integrity	
		Sodium laurethsulfate (SLES)	Improves skin dryness	
	Cationic surfactant	Cetrimonium chloride	Improved skin moisturisation	[41]
	Nonionic surfactant	Polysorbate 20	It helps to stabilise the formulation	[43]
		Polysorbate 80	Acts as an emulsifier for blending oils	
	Amphoteric surfactant	Cocamidopropyl hydroxysultaine (CHSB)	It is a foam-boosting agent and viscosity builder	[45]
		Disodium cocoamphodiacetate (DSCA)	It has mild cleansing activity and is used in conditioners for hair softening	

DOI: http://dx.doi.org/10.5772/intechopen.112925

S.no	Category	Ingredients	Use/Comment	Reference
4.	Preservatives			[47]
	Natural preservatives			
	Plant-based Essential oil	Rosemary extract	Exhibit excellent antioxidant properties	
		Tea tree oil	They have antimicrobial properties	
		Lavender oil Clove oil	Effective against gram-positive and gram-negative bacteria	
	Synthetic Preservatives			[51] [52] [53] [55] [56]
	Parabens	Methylparaben	It is an anti-fungal agent often used in various cosmetics and personal care products	
		Propylparaben	A non-volatile compound used as an antimicrobial preservative	
	Formaldehyde-releasing agents	Diazolidinyl urea	Effective preservative against yeast and moulds	
		Imidazolidinyl urea	It is an antimicrobial agent used in personal care products	
	Quaternary ammonium compounds	Benzalkonium chloride	Used in hand sanitiser to reduce *S. aureus* contamination	
		Cetrimonium chloride	Used as topical antiseptic and preservative	
	Organic acids	Benzoic acid	It is used as an antimicrobial preservative	
		Salicylic acid	It is used as an exfoliator and has antiseptic properties	
		Phenoxyethanol	Used as a stabiliser in perfumes and soaps	
		Triclosan	Used as an antimicrobial preservative in shampoos	
		Chlorhexidine	A study suggested chlorhexidine-containing toothpaste or mouthwashes to improve symptoms and resolution of gingivitis	
	Others			
5.	Antioxidants	Vitamin E	Increase skin hydration, make the	[60] [61]
		Vitamin C	Makes skin shiny and decrease sebum production	
		Green tea extract	Antimicrobial properties reduce skin redness and prevent wrinkles	
6.	UV filters	Zincoxide Titaniumdioxide	Both of these are UV rays absorber and offer sunscreen transparency when microsized	[64]

Table 1.
Common cosmetic ingredients and their use.

could be defined as "any cosmetic product containing UV filters in its formulation in order to protect the skin from the solar deleterious UV light, avoiding or minimizing the damage that this radiation might cause on human health."

Since the finished product's UV filter concentration is connected to its sun protection effectiveness, which is often represented by the labelled sun protection factor (SPF), analytical monitoring of sunscreen cosmetics is required. In order to assure safety, it is also vital to ensure that the concentration levels are lower than those authorised by law, as applying sunscreen has a number of recognised unfavourable dermatological side effects [62, 63]. A study on the safety and effectiveness of titanium and zinc oxide nanoparticles suggested that the replacement of microsized TiO_2 and ZnO particles by nanoparticles offers the aesthetically desired sunscreen transparency but at the sacrifice of wide ultraviolet A rays protection. Common cosmetic ingredients and their use are summarized in **Table 1** [64].

4. How ingredients are selected for cosmetics

Millions of consumers use PCPs (personal care products) and the chemicals they contain daily. Personal care products provide local (skin and eye) exposure and are utilised in the oral cavity, on the face, lips, eyes and mucosa, even though human external contact with a drug seldom causes its penetration through the skin and considerable systemic exposure [65]. The safety of personal care products and their constituents has received more attention in recent years; as a result, the toxicological safety evaluation of these products is a relatively new field that developed in the second half of the 20th century [66]. In addition to the effectiveness of the active ingredient(s), the success of a cosmetic formulation also depends on customer acceptability, which is greatly influenced by the product's sensory qualities. As a result, several research have concentrated on sensory assessments of cosmetics in order to formulate solutions that offer sufficient effectiveness and appealing aesthetic qualities, thereby satisfying customers' demands [67]. In most countries, when a cosmetic ingredient is proven to be hazardous, the regulatory agency prohibits its use. The proof of a cosmetic hazardous effect requires robust scientific evidence from several studies performed in vitro, in animals as well in humans. Moreover, toxicity tests should be performed using the ingredient in the concentration used in cosmetics and in the condition of use (e.g., route of administration, acute or chronic exposure) [68].

All parties involved—consumers, regulatory bodies and producers—have the same goal: they want cosmetics/personal care products (PCP) and the components in them to be safe, meaning they should pose no or very little health risk to the user during regular usage. On the basis of in vitro skin penetration results, the size of possible human systemic exposure may be approximated; however, the test's limitations should be considered [65].

Natural and safe products are more in demand from consumers, and as a result, there is a significant increase in the quantity of waste produced by industrial operations. Various tests, like the microbial limit guidelines, are carried out to specify the degree of microbiological contamination in non-sterile products like cosmetic formulations. By including an appropriate preservative in the goods that ensures the control of microbial growth even before they are sold, these values should be maintained in the products during their use despite the unavoidable contamination by users. Numerous industries produce different sorts of disposable byproducts with valuable chemicals. The prospective and existing uses of chemicals and extracts derived from agronomical disposable wastes in the beauty industry were the focus of a study published in 2015. There are numerous instances of active compounds in cosmetics that come from dairy, meat and fish. These goods offer a viable alternative to the ordinary plant-derived extracts that are more frequently used in cosmetic formulations because they are efficient, affordable and biodegradable. Byproduct extracts, or those derived from the processing of fruits and vegetables, are a viable "green" substitute for the conventional plant-derived extracts that are frequently used in cosmetics. The most processed tropical fruit, pineapple (Ananascomosus), produces a lot of byproducts [69]. According to research, the amount of L-ascorbic acid in the pineapple rind was higher than in the fruit as a whole, and it was demonstrated that UV-C radiation boosted the L-ascorbic acid concentration in the rind [70].

5. Innovations in cosmetic ingredients

It is challenging to create cosmetics from entirely natural raw ingredients. The difficulty is in choosing components that may be justifiably described as "natural" and combining them to create cosmetics that operate similarly to those made of synthetic elements. In an effort to avoid health risks and lessen pollution, many people prefer "green cosmetics," or ecologically friendly lotions, makeup and beauty items. Furthermore, the COVID-19 pandemic's frequent mini-lockdowns have increased awareness of the connection between exterior and internal well-being and physical beauty. In turn, this has led to a reduction in makeup preferences and an increase in preferences for skincare products. The new market demands are being answered by nutricosmetics, which combines the advantages of cosmetic procedures with the advantages of dietary supplementation to enhance the beauty of our bodies [71].

According to a study by Macroalgae (seaweeds), which are divided into three families based on their predominant pigment: Rhodophyceae (red algae), Chlorophyceae (green algae), and Pheophyceae (brown algae), natural or nature-based ingredients are now frequently used to create new cosmetic products. Several macroalgal extracts and bioactive substances have shown promise in treating various skin diseases [72].

Nanocarrier technology has successfully overcome traditional methods' limitations, enabling the development of advanced drug delivery systems [73]. Christian Dior originally introduced liposome-based cosmetics (Capture) in 1986; soon after that, functional cosmetics using nanocarrier technology were developed [74].

In cosmetics, nanoemulsion is frequently used topically, mostly for hydrating and moisturising skin. This method has gained popularity because of its simplicity in formulation, handling and manufacturing. In a study of the in vitro and in vivo safety evaluations of nanoemulsion as a skin moisturiser, a study published in 2022 found that the product had improved penetration and absorption [75].

The morphologies and chemical makeup of the nanoparticles vary. Nevertheless, they are utilised in physical UV filters and sunscreen preparations (such as TiO2-, ZnO-, CeO2- and ZrO2-nanoparticles) [76]. As thickeners, silica and clay nanoparticles are also included. In addition, silica and clay nanoparticles are added as thickeners [77]. In a study conducted in 2008, investigators created flexible liposomes for topical application in cosmetics. They discovered that due to their chemical nature, flexible liposomes are the only option for carrying active molecules into the deeper skin layers (biological syringe). They also act as pharmaceutical or cosmetic ingredients [78]. Non-phospholipidoligolamellar lipid vesicles with a diameter of 0.1 to 1.0

microns that are a variant of liposomes or modified niosomes are another innovative formulation of novasomes. They are created by mixing the monoester of polyoxyethylene fatty acids, cholesterol, and free fatty acids in a ratio of 74/22/4. By having the ability to adhere to skin or hair shafts, they provide even more excellence for use in cosmetic preparations. The effectiveness and texture of these cosmetics are improved, and sustained release is also made possible [79].

Consumers and experts are concerned about clean beauty, a trend in cosmetics that utilises less synthetic components and more plant-based and herbal compounds. The scientific community faces a significant dilemma as a result of PCPs' ability to expose users to hazardous substances. Women use an average of eight items per day (some report using up to 30 products per day), giving their bodies a daily dose of a variety of substances that may interact harmoniously or antagonistically. Endocrine disruption, cancer, damage to the reproductive system, and delays in children's neurodevelopment have all been associated with exposure to these substances [80]. In a recent development, investigators looked into the capacity of silk fibroin nanoparticles (SFNs) to transport, encapsulate and heat-protect the phenolic components of the ethanolic extract of guava leaves. The findings demonstrated that the guava extract nearly completely lost all of its antioxidant capacity after being exposed to a high temperature of 70°C for 24 hours, but the extract-loaded SFNs were able to retain the extract activity. The efficacy and safety of cosmetics have improved as a result of all these nanoformulations [81].

The dangers that hazardous substances to consumers are known to pose to manufacturers and merchants. This is true in part as a result of the rise in consumer advocacy and the clean beauty movement, which have brought attention to the need for PCPs made with safer, cleaner ingredients (excluding toxic chemicals that have a negative impact on human health) and transparent labelling (excluding the umbrella term "fragrance," which can contain a variety of harmful, unregulated chemicals and misleading labels with words such as "natural," "organic," and "eco"). Some businesses, like Sephora and others, have started promoting and highlighting specific branded items as being clean, ecologically sustainable, or planet-positive in an effort to assist shoppers [82].

6. Conclusion

The market for cosmetic or personal care goods has recently grown significantly. Environmentally friendly and herbal products are being used increasingly frequently. Bacterial biosurfactants (BS), a biocompatible, low-toxic alternative surfactant, are becoming more and more common in the manufacturing of industrial products. Chemical surfactants should be replaced with less damaging chemicals for the health of your skin since they might cause allergic reactions and skin irritations. This chapter provides considerable insights about cosmetics and how the role of ingredients can change a cosmetic formulation entirely. It also addresses how a large population began using cosmetic products with synthetic ingredients for their quick effects, which had some benefits like being quicker to apply, easier to store and more attractive, as well as some drawbacks like sporadic deterioration, more unwanted side effects, skin allergies and cost-effectiveness. However, the majority of the population in recent eras has moved towards clean beauty alternatives, with the changing trends in cosmetics ingredients. Plant-based products are gaining much more importance than synthetic and conventional ingredients. Since India is the source of the updated systems of traditional medicine like Siddha, Ayurveda and Unani and is rich in flora, plants have been employed for simple cures since ancient times and have gained popularity. Several commercial skincare formulas are available for skin whitening, UV protection and anti-ageing. These readily available herbal preparations offer skin care advantages in addition to therapeutic ones. The administration of herbal cosmetically active components requires a new study direction since skin condition research has become highly fascinating and has achieved considerable advancements. In particular, nanoformulation is the field that, when investigated, will lead to much more breakthroughs in cosmetics and personal care goods. Cosmetic products will have a very bright future in the next years.

Conflict of interest

Authors declare no conflict of interest.

Author details

Shivangi Sharma[1], Usama Ahmad[2*], Juber Akhtar[2], Anas Islam[2], Mohd Muazzam Khan[2] and Naureen Rizvi[2]

1 Institute of Pharmacy, Shri Ramswaroop Memorial University, Barabanki, India

2 Faculty of Pharmacy, Integral University, Lucknow, India

*Address all correspondence to: usamaahmad.10@outlook.com

References

[1] Mitsui T, editor. Cosmetics and skin. New Cosmetic Science. Elsevier; 1997

[2] Parish LC, Crissey JT. Cosmetics: A historical review. Clinics in Dermatology. 1988;**6**(3):1-4

[3] Montenegro L, Paolino D, Puglisi G. Effects of silicone emulsifiers on in vitro skin permeation of sunscreens from cosmetic emulsions. Journal of Cosmetic Science. 2004;**55**(6):509-518

[4] Gawade R, P, Chinke SL, Alegaonkar PS. Chapter 17 - Polymers in Cosmetics, Polymer Science and Innovative Applications. Elsevier; 2020. pp. 545-565. DOI: 10.1016/B978-0-12-816808-0.00017-2

[5] AOCS. Your globalfat oil connection Magazine article [Internet]. Available from: https://www.aocs.org/stay-informed/inform-magazine/featured-articles/an-introduction-to-cosmetic-technology-april-2015?SSO=True [Accessed 2023-4-25]

[6] Lochhead RY. The role of polymers in cosmetics: Recent trends. American Chemical Society; 2007

[7] Patil A, Ferritto MS. Polymers for Personal Care and Cosmetics. ACS Symposium Series. Vol. 1148. American Chemical Society. 2013. ISBN: 9780841229051. eISBN: 9780841229068

[8] Bhattacharya B, Ghosh TK, Das N. Application of bio-surfactants in cosmetics and pharmaceutical industry. Scholars Academic Journal Pharmacy. 2017;**6**(7):320-329. DOI: 10.21276/sajp

[9] Desai JD, Banat IM. Microbial production of surfactants and their commercial potential. Microbiology and Molecular Biology Reviews. 1997;**61**(1):47-64. DOI: 10.1128/mmbr.61.1.47-64.1997

[10] Rodrigues L, Banat IM, Teixeira J, Oliveira R. Biosurfactants: Potential applications in medicine. Journal of Antimicrobial Chemotherapy. 2006;**57**(4):609-618. DOI: 10.1093/jac/dkl024

[11] Moravkova T, Filip P. The influence of emulsifier on rheological and sensory properties of cosmetic lotions. Advances in Materials Science and Engineering. 2013;**2013**. DOI: 10.1155/2013/168503

[12] Hahn HJ, Jung HJ, Schrammek-Drusios MC, Lee SN, Kim JH, Kwon SB, et al. Instrumental evaluation of anti-aging effects of cosmetic formulations containing palmitoyl peptides, Silybummarianum seed oil, vitamin E and other functional ingredients on aged human skin. Experimental and Therapeutic Medicine. 2016;**12**(2):1171-1176. DOI: 10.3892/etm.2016.3447

[13] Draelos ZD, Gunt H, Zeichner J, Levy S. Clinical evaluation of a nature-based Bakuchiol anti-aging moisturizer for sensitive skin. Journal of Drugs in Dermatology: JDD. 2020;**19**(12):1181-1183. DOI: 10.36849/jdd.2020.5522

[14] Diffey BL, Tanner PR, Matts PJ, Nash JF. In vitro assessment of the broad-spectrum ultraviolet protection of sunscreen products. Journal of the American Academy of Dermatology. 2000;**43**(6):1024-1035. DOI: 10.1067/mjd.2000.109291

[15] Gordobil O, Olaizola P, Banales JM, Labidi J. Lignins from agroindustrial byproducts as natural ingredients

for cosmetics: Chemical structure and in vitro sunscreen and cytotoxic activities. Molecules. 2020;**25**(5):1131. DOI: 10.3390/molecules25051131

[16] Nunes MA, Rodrigues F, Oliveira MB. Grape processing byproducts as active ingredients for cosmetic proposes. In: Handbook of Grape Processing Byproducts. Academic Press.; 2017. pp. 267-292. DOI: 10.1016/B978-0-12-809870-7.00011-9

[17] Camargo FB Jr, Gaspar LR, Maia Campos PM. Skin moisturizing effects of panthenol-based formulations. Journal of Cosmetic Science. 2011;**62**(4):361

[18] Censi R, Vargas Peregrina D, Lacava G, Agas D, Lupidi G, Sabbieti MG, et al. Cosmetic formulation based on an Açai extract. Cosmetics. 2018;**5**(3):48

[19] Marto J, Neves Â, Gonçalves LM, Pinto P, Almeida C, Simões S. Rice water: A traditional ingredient with anti-aging efficacy. Cosmetics. 2018;**5**(2):26

[20] Kala SL, Palaparthi S. Formulation and invitro evaluation of poly herbal anti aging face cream. World Journal of Pharamaceutical Research. 2017;**6**(13):713

[21] Igbokwe NH, Idowu AO, Abdulwahab UB, Oseni BA, Isibor K. Antibacterial profiling of methanolic leaf extracts and herbal cosmetic cream formulations containing the leaf extracts of Urticadioica, Amaranthusviridis and Aloe vera

[22] Kusrini E, Mawarni DP, Wulandari DA, Ayuningtyas K, Usman A. Formulation and characterization of lip balm made from beeswax, almond oil, virgin coconut oil and honey. In: AIP Conference Proceedings 2020. AIP Publishing LLC;

[23] Djuris J, Vasiljevic D, Jokic S, Ibric S. Application of D-optimal experimental design method to optimize the formulation of O/W cosmetic emulsions. International Journal of Cosmetic Science. 2014;**36**(1):79-87

[24] Khan H, Akhtar N, Mahmood T, Jameel A, Mohsin S. Preliminary 1 month stability screening of cosmetic multiple emulsions (W/O/W) prepared using cetyldimethiconecopolyol and Polysorbate 80. International Journal of Cosmetic Science. 2015;**37**(1):76-81

[25] Martínez de MurguíaFernández L, PuigAlgora G, BajonaRoig M, Bacchini G. Effectiveness and tolerability of a squalane and dimethicone-based treatment for head lice. Parasitology Research. 2021;**120**(5):1883-1890

[26] Morrison DS, Schmidt J, Paulli R. The scope of mineral oil in personal care products and its role in cosmetic formulation. Ammonio Lattato Attivato. 1996;**14**:111-118

[27] Kurul E, Hekimoğlu S. Skin permeation of two different benzophenone derivatives from various vehicles. International Journal of Cosmetic Science. 2001;**23**(4):211-218

[28] Mukherjee S, Edmunds M, Lei X, Ottaviani MF, Ananthapadmanabhan KP, Turro NJ. Original contribution: Stearic acid delivery to corneum from a mild and moisturizing cleanser. Journal of Cosmetic Dermatology. 2010;**9**(3):202-210

[29] Montenegro L, Carbone C, Paolino D, Drago R, Stancampiano AH, Puglisi G. In vitro skin permeation of sunscreen agents from O/W emulsions. International Journal of Cosmetic Science. 2008;**30**(1):57-65

[30] Guaratini T, Gianeti MD, Campos PM. Stability of cosmetic

formulations containing esters of vitamins E and a: Chemical and physical aspects. International Journal of Pharmaceutics. 2006;**327**(1-2):12-16. DOI: 10.1016/j.ijpharm.2006.07.015

[31] Leite MG, Campos PM. Development and efficacy evaluation of hair care formulations containing vegetable oils and silicone. International Journal of Phytocosmetics and Natural Ingredients. 2018;**5**(1):9. DOI: 10.15171/ijpni.2018.09

[32] Andre P. Hyaluronic acid and its use as a "rejuvenation" agent in cosmetic dermatology. In: Seminars in Cutaneous Medicine and Surgery. 2004. pp. 218-222. DOI: 10.1016/j.sder.2004.09.002

[33] Sakulwech S, Lourith N, Ruktanonchai U, Kanlayavattanakul M. Preparation and characterization of nanoparticles from quaternizedcyclodextrin-grafted chitosan associated with hyaluronic acid for cosmetics. Asian Journal of Pharmaceutical Sciences. 2018;**13**(5):498-504. DOI: 10.1016/j.ajps.2018.05.006

[34] Sionkowska A, Kaczmarek B, Michalska M, Lewandowska K, Grabska S. Preparation and characterization of collagen/chitosan/hyaluronic acid thin films for application in hair care cosmetics. Pure and Applied Chemistry. 2017;**89**(12):1829-1839. DOI: 10.1515/pac-2017-0314

[35] Pounikar YO, Jain PU, Khurana NA, Omray LK, Patil S, Gajbhiye A. Formulation and characterization of Aloe vera cosmetic herbal hydrogel. International Journal of Pharmacy and Pharmaceutical Sciences. 2012;**4**(4):85-86

[36] Cserháti T, Forgács E, Oros G. Biological activity and environmental impact of anionic surfactants. Environment International. 2002;**28**(5):337-348. DOI: 10.1016/S0160-4120(02)00032-6

[37] Leoty-Okombi S, Gillaizeau F, Leuillet S, Douillard B, Le Fresne-Languille S, Carton T, et al. Effect of sodium lauryl sulfate (SLS) applied as a patch on human skin physiology and itsmicrobiota. Cosmetics. 2021;**8**(1):6. DOI: 10.3390/cosmetics8010006

[38] Bujak T, Nizioł-Łukaszewska Z, Wasilewski T. Sodium lauryl sulfate vs. sodium coco sulfate. Study of the safety of use anionic surfactants with respect to their interaction with the skin. Tenside Surfactants Detergents. 2019;**56**(2):126-133. DOI: 10.3139/113.110599

[39] Burnett CL, Bergfeld WF, Belsito DV, Hill RA, Klaassen CD, Liebler D, et al. Final report of the cosmetic ingredient review expert panel on the safety assessment of cocamidopropyl betaine (CAPB). International Journal of Toxicology. Aug 2012;**31**(4). DOI: 10.1177/1091581812447202

[40] Li LF. A study of the sensitization rate to cocamidopropylbetaine in patients patch tested in a university hospital of Beijing. Contact Dermatitis. 2008;**58**(1):24-27. DOI: 10.1111/j.1600-0536.2007.01251.x

[41] Herrwerth S, Leidreiter H, Wenk HH, Farwick M, Ulrich-Brehm I, Grüning B. Highly concentrated cocamidopropylbetaine–the latest developments for improved sustainability and enhanced skin care. Tenside Surfactants Detergents. 2008;**45**(6):304-308. DOI: 10.3139/113.100387

[42] Hunter JE, Fowler JF Jr. Safety to human skin of cocamidopropylbetaine: A mild surfactant for personal-care products. Journal of Surfactants

and Detergents. 1998;**1**(2):235-239. DOI: 10.1007/s11743-998-0025-3

[43] Jang HJ, Shin CY, Kim KB. Safety evaluation of polyethylene glycol (PEG) compounds for cosmetic use. Toxicological Research. 2015;**31**:105-136. DOI: 10.1016/j.tox.2005.06.001

[44] Bhoyrul B, Solman L, Kirk S, Orton D, Wilkinson M. Patch testing with alkyl glucosides: Concomitant reactions are common but not ubiquitous. Contact Dermatitis. 2019;**80**(5):286-290. DOI: 10.1111/cod.13186

[45] Baranova I, Zaika S, Bezpala Y, Roik O, Zaporozhska S, Shostak L. Development of foaming shampoo base for the treatment of Seborrheic dermatitis. Journal of Advanced Pharmacy Education and Research. 2020;**10**(1):143-149

[46] Halla N, Fernandes IP, Heleno SA, Costa P, Boucherit-Otmani Z, Boucherit K, et al. Cosmetics preservation: A review on present strategies. Molecules. 2018;**23**(7):1571. DOI: 10.3390/molecules23071571

[47] Soto ML, Falqué E, Domínguez H. Relevance of natural phenolics from grape and derivative products in the formulation of cosmetics. Cosmetics. 2015;**2**(3):259-276. DOI: 10.3390/cosmetics2030259

[48] Kunicka-Styczyńska A, Sikora M, Kalemba D. Antimicrobial activity of lavender, tea tree and lemon oils in cosmetic preservative systems. Journal of Applied Microbiology. 2009;**107**(6):1903-1911. DOI: 10.1111/j.1365-2672.2009.04372.x

[49] Dreger M, Wielgus K. Application of essential oils as natural cosmetic preservatives. Herbapolonica. 2013;**59**(4):142-156. DOI: 10.2478/hepo-2013-0030

[50] Pedersen S, Marra F, Nicoli S, Santi P. In vitro skin permeation and retention of parabens from cosmetic formulations. International Journal of Cosmetic Science. 2007;**29**(5):361-367. DOI: 10.1111/j.1468-2494.2007.00388.x

[51] Monakhova YB, Kuballa T, Mildau G, Kratz E, Keck-Wilhelm A, Tschiersch C, et al. Formaldehyde in hair straightening products: Rapid 1H NMR determination and risk assessment. International Journal of Cosmetic Science. 2013;**35**(2):201-206. DOI: 10.1111/ics.12027

[52] Lv C, Hou J, Xie W, Cheng H. Investigation on formaldehyde release from preservatives in cosmetics. International Journal of Cosmetic Science. 2015;**37**(5):474-478. DOI: 10.1111/ics.12212

[53] Bondurant S, McKinney T, Bondurant L, Fitzpatrick L. Evaluation of a benzalkonium chloride hand sanitizer in reducing transient Staphylococcus aureus bacterial skin contamination in health care workers. American Journal of Infection Control. 2020;**48**(5):522-526. DOI: 10.1016/j.ajic.2019.08.030

[54] Kligman D, Kligman AM. Salicylic acid peels for the treatment of photoaging. Dermatologic Surgery. 1998;**24**(3):325-328. DOI: 10.1111/j.1524-4725.1998.tb04162.x

[55] Arif T. Salicylic acid as a peeling agent: A comprehensive review. Clinical, Cosmetic and Investigational Dermatology. 2015;**26**:455-461. DOI: 10.2147/CCID.S84765

[56] Magaz VR, Llovera BF, Martí M, Garre A. Clinical impact and cosmetic acceptability of chlorhexidine-enriched toothpaste and mouthwash application on periodontal disease: A randomized clinical study. The Journal

of Contemporary Dental Practice. 2018;**19**(11):1295-1300

[57] Lupo MP. Antioxidants and vitamins in cosmetics. Clinics in Dermatology. 2001;**19**(4):467-473. DOI: 10.1016/s0738-081x(01)00188-2

[58] Darr D, Combs S, Dunston S, et al. Topical vitamin C protects porcine skin from ultraviolet radiation-induced damage. The British Journal of Dermatology. 1992;**127**:247-253

[59] Chan AC. Partners in defense, vitamin E and vitamin C. Canadian Journal of Physiology and Pharmacology. 1993;**71**:725-731

[60] Akhtar N, Yazan Y. Formulation and in-vivo evaluation of a cosmetic multiple emulsion containing vitamin C and wheat protein. Pakistan Journal of Pharmaceutical Sciences. 2008;**21**(1):45-50

[61] Dingler A, Blum RP, Niehus H, Muller RH, Gohla S. Solid lipid nanoparticles (SLNTM/LipopearlsTM) a pharmaceutical and cosmetic carrier for the application of vitamin E in dermal products. Journal of Microencapsulation. 1999;**16**(6):751-767. DOI: 10.1080/026520499288690

[62] Funk JO, Dromgoole SH, Maibach HL. Dermatologic Clinics. 1995;**13**:473

[63] Schauder S, Ippen H. Contact and photocontact sensitivity to sunscreens: Review of a 15-year experience and of the literature. Contact Dermatitis. Nov 1997;**37**(5):221-232

[64] Smijs TG, Pavel S. Titanium dioxide and zinc oxide nanoparticles in sunscreens: Focus on their safety and effectiveness. Nanotechnology, Science and Applications. 2011;**4**:95-112. DOI: 10.2147/NSA.S19419

[65] Nohynek GJ, Antignac E, Re T, Toutain H. Safety assessment of personal care products/cosmetics and their ingredients. Toxicology and Applied Pharmacology. 2010;**243**(2):239-259. DOI: 10.1016/j.taap.2009.12.001

[66] Draize JH. Methods for study of irritation and toxicity of substances applied topically to the skin and mucous membranes. Journal of Pharmacology and Experimental Therapeutics. 1944;**82**:377-390

[67] Garrigue JL, Ballantyne M, Kumaravel T, Lloyd M, Nohynek GJ, Kirkland D, et al. In vitro genotoxicity of Para-phenylenediamine and its Nmonoacetyl and N,N'-diacetyl metabolites. Mutation Research. 2006;**608**:58-71

[68] Cosmetic Ingredient Review (CIR). Guidelines for the Safety Assessment of Cosmetic Ingredients. Washington, DC: Cosmetic Ingredient Review; 2023

[69] Barbulova A, Colucci G, Apone F. New trends in cosmetics: Byproducts of plant origin and their potential use as cosmetic active ingredients. Cosmetics. 2015;**2**(2):82-92. DOI: 10.3390/cosmetics2020082

[70] Freitas A, Moldao-Martins M, Costa H, Albuquerque T, Valentea A, Sanches-Silvaa A. Effect of UV-C radiation on bioactive compounds of pineapple (Ananas. Comosus L. Merr.) byproducts. Journal of the Science of Food and Agriculture. 2015;**95**:44-52

[71] Dini I, Laneri S. The new challenge of green cosmetics: Natural food ingredients for cosmetic formulations. Molecules. 2021;**26**(13):3921. DOI: 10.3390/molecules26133921

[72] Juliano C, Magrini GA. Cosmetic functional ingredients from botanical

sources for anti-pollution skincare products. Cosmetics. 2018;**5**(1):19. DOI: 10.3390/cosmetics5010019

[73] Zhou H, Luo D, Chen D, Tan X, Bai X, Liu Z, et al. Current advances of nanocarrier technology-based active cosmetic ingredients for beauty applications. Clinical, Cosmetic and Investigational Dermatology. 2021;**14**:867-887. DOI: 10.2147/CCID.S313429

[74] Müller-Goymann CC. Physicochemical characterization of colloidal drug delivery systems such as reverse micelles, vesicles, liquid crystals and nanoparticles for topical administration. European Journal of Pharmaceutics and Biopharmaceutics. 2004;**58**(2):343-356. DOI: 10.1016/j.ejpb.2004.03.028

[75] Mohd-Nasir H, Aziz ZA, Setapar SH. In vitro and in vivo safety evaluation of nanoemulsion as skin moisturizer. Nanotechnology for the Preparation of Cosmetics Using Plant-Based Extracts. 2022:345-354. DOI: 10.1016/B978-0-12-822967-5.00012-6

[76] Sumit K, Vivek S, Sujata S, Ashish B. Herbal cosmetics: Used for skin and hair. Inventi Journals. 2012;**2012**:1-7

[77] Choy J-H, Choi S-J, Oh J-M, Park T. Clay minerals and layered double hydroxides for novel biological applications. Applied Clay Science. 2007;**36**:122-132

[78] Blume G. Flexible liposomes for topical applications in cosmetics. Science and Applications of Skin Delivery Systems. 2008:269-282

[79] Singh A, Malviya R, Sharma PK. Novasome-a breakthrough in pharmaceutical technology a review article. Advances in Biology Research. 2011;**5**:184-189

[80] Dodson RE, Cardona B, Zota AR, Robinson Flint J, Navarro S, Shamasunder B. Personal care product use among diverse women in California: Taking stock study. Journal of Exposure Science & Environmental Epidemiology. 2021;**31**(3):487-502. DOI: 10.1038/s41370-021-00327-3

[81] Pham DT, Nguyen DX, Lieu R, Huynh QC, Nguyen NY, Quyen TT, et al. Silk nanoparticles for the protection and delivery of guava leaf (Psidiumguajava L.) extract for cosmetic industry, a new approach for an old herb. Drug Delivery. 2023;**30**(1):2168793. DOI: 10.1080/10717544.2023.2168793

[82] McDonald JA, Llanos AA, Morton T, Zota AR. The environmental injustice of beauty products: Toward clean and equitable beauty. American Journal of Public Health. 2022;**112**(1):50-53

Chapter 2

Skin Care Nanocosmetics

Júlia Scherer Santos, Carolina Sousa Ponciano, Thaís Nogueira Barradas, Mirsiane Pascoal Costa and Guilherme Diniz Tavares

Abstract

The improvement of stability, solubility, spreadability and skin penetration of cosmetics as well as the reduction of oxidation may be achieved by nanocarriers. In that regard, many cosmetic industries have launched nanocosmetics due to their performance improvement. Sunscreens, moisturizers and anti-aging products based on nanotechnology are available worldwide. In addition, vegetable extracts loaded into nanocarriers have also been employed as a strategy to increase their skin penetration. In this chapter, the main contributions of polymeric nanocarriers, lipid nanocarriers and vesicular nanocarriers to skin care cosmetics were approached.

Keywords: nanocosmetics, polymeric carriers, lipid carriers, vesicular carriers, skin

1. Introduction

Since the beginning of human civilization, there is a search for beauty and well-being [1]. Currently, people are willing to perform different procedures to prevent skin aging. This trend is reflected in an increased number of skin care products over the years [2]. Along with population aging and increased cosmetic consumption, innovative cosmetics have been targeted by cosmetic industries to develop improved performance products [3]. In this context, nanocosmetics are reported since 1986 [4] to improve formulations efficacy and stability [4, 5].

Several cosmetics based on nanotechnology are available for skin, hair and skin appendages. Among these, skin care nanocosmetics account for most commercial products, with emphasis on those with anti-aging, moisturizing, anti-wrinkle, antioxidant and sunscreen properties [6]. Mainly for women, anti-aging and anti-wrinkle claims products are highlighted [7].

This chapter will address the state of art and the most recent contributions of nanocosmetics for skin care, with emphasis on the most employed nanocarriers. Hence, polymeric carriers (i.e., nanocapsules and nanospheres), lipid carriers (i.e., solid lipid nanoparticles, nanostructured lipid carriers and nanoemulsions) and vesicular carriers (i.e., liposomes, transfersomes and niosomes) will be approached.

2. Cosmetics based on polymeric nanocarriers

Polymeric nanoparticles are bioactive carrier systems formed essentially by polymers of synthetic or natural origin. Among these, the most widely used are poly (lactic acid) (PLA), poly (d,l-lactic acid-co-glycolic acid) (PLGA), poly(alkyl cyanoacrylate) (PACA), and poly (ε-caprolactone) (PCL) (synthetic polymers) and gelatin, albumin and chitosan (natural polymers) [8]. These nanoparticles have been investigated to increase the stability of cosmetic actives, control their release, increase cutaneous penetration, and avoid incompatibilities between formulation ingredients [4].

These systems can be divided into nanocapsules and nanospheres (**Figure 1**), which differ according to their composition and structural organization. Nanocapsules (**Figure 1b**) consist of a polymeric shell arranged around an oily core, and the active ingredient may be dissolved in this core and/or adsorbed to the polymeric wall. On the other hand, nanospheres (**Figure 1a**), which do not have oil in their composition, are formed by a polymeric matrix, where active ingredient is uniformly dispersed or solubilized inside this matrix and may be retained or adsorbed [9].

Nanocapsules have been extensively investigated as vehicles for chemicals sunscreen such as octyl methoxycinnamate, octyl salicylate, and benzophenone-3 [10, 11]. From this perspective, there are reports of the development of polymeric nanocapsules coated with chitosan, which is a cationic, biocompatible, bioadhesive and FDA-approved polymer, to increase the skin adhesion and the photoprotective effect of sunscreen [12]. Also, these nanoparticles are able to form a protective film on the skin surface and control the penetration and permeation of the encapsulated substances [13]. In that regard, *ex vivo* permeation studies performed on porcine ear skin have decreased benzophenone-3 permeation [14]. Another promising application of polymeric nanocapsules in skin care is the increased anti-acne activity of tea tree oil [15].

As to nanospheres, they are used to encapsulate fragrances and vitamins [16]. In this sense, there are reports that fragrances encapsulated in nanospheres remained on the skin after a long period of application. In addition, PLGA nanospheres containing ascorbyl tetraisopalmitate had a higher skin deposition [17]. Apart from that, chitosan content affected the encapsulation efficiency of alpha-arbutin loaded in chitosan nanoparticles. Also, the mode of incorporation of the drug also influenced the encapsulation efficiency. Therefore, formulation optimization is essential to obtain suitable physicochemical features and an improvement of alpha-arbutin-loaded chitosan nanoparticles on melasma treatment [18].

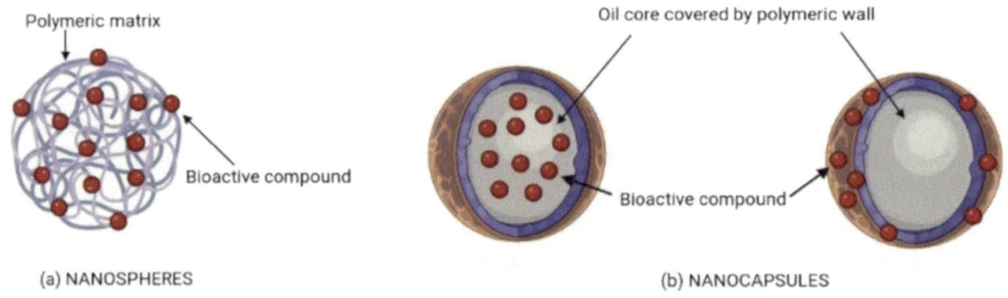

Figure 1.
Nanospheres and nanocapsule structure [9].

Regarding nanospheres' application to photoprotection, gelatin nanoparticles were applied for encapsulation of natural compounds, such as rutin [19] and *Baccharis antioquensis* extract [20]. In both reports, there was improved photostability and increased photoprotection [19, 20]. **Table 1** summarizes the main results of polymeric nanoparticles and other nanocarriers applied to cosmetics.

3. Cosmetics based on lipid nanocarriers

Lipid-based nanocarriers are composed of a lipid core, where they encapsulate one or more biologically active substances. In general, they can provide less toxicity, in addition to several other favorable attributes, such as increased penetration into the skin, possibility of sustained release and protection of actives against degradation, among others [3, 36]. Lipid-based nanocarriers comprise solid lipid nanoparticles (SLNs), nanoemulsions (NE) and nanostructured lipid carriers (NLCs). Lipid-based nanocarriers are increasingly present in the cosmetics market in skincare products, due to their beneficial properties for the skin [37]. They are able to effectively solubilize and deliver hydrophobic active ingredients to the skin, resulting in greater efficacy of cosmetic products [36].

SLNs comprise the first generation of lipid-based nanocarriers, being introduced in the last decade of the twentieth century. SLNs are constituted of a solid lipid core stabilized by a surfactant layer, being able to successfully deliver lipophilic actives to the upper skin layers [38, 39]. In practical terms, SLNs offer the possibility of large-scale production, not relying on organic solvents for their preparation, and rigid morphology, which increase their stability [39, 40]. SLNs are considered the most

Nanocarriers		Cosmetic ingredient	Main outcomes	Cosmetic use	Ref.
Polymeric nanocarriers	Nanocapsules	BP-3*	Decreased permeation	Sunscreen	[14]
		bp-3*	Increased skin adhesion Increased photoprotection	Sunscreen	[12]
		Tea tree oil	Increased anti-acne activity	Anti-acne	[15]
	Nanospheres	Rutin EHDP* EHMC* BMDBM*	Increased photostability Increased sunscreen protection factor	Sunscreen Antioxidant	[19]
		VC-IP*	Increased permeation	Antioxidant	[17]
Lipid nanocarriers	Solid lipid nanoparticles	Caffeine	Faster skin permeation	Anti-aging	[21]
		Oxybenzone	Increased skin hydration Increased photoprotection	Sunscreen	[22]
	Nanostructured lipid carriers	Coenzyme Q10	Improved skin penetration	Anti-aging Antioxidant	[23]
		NDAG*	Prolonged release Higher antioxidant ability	Antioxidant	[24]
		Carrot extract Marigold extract	Increased skin hydration Increased skin elasticity	Antioxidant	[25]

Nanocarriers		Cosmetic ingredient	Main outcomes	Cosmetic use	Ref.
Vesicular nanocarriers	Liposomes	Vitamin C	Higher photostability	Antioxidant	[26]
		Lactobacillus rhamnosus	Lower cytotoxicity	Antioxidant Antimicrobial	[27]
		4nBR* RSV*	Decreased melasma index	Hyperpigmentation	[28]
		Cysteamine	Lower oxidation	Hyperpigmentation	[29]
		Ascorbic acid	Increased skin permeation	Antioxidant Anti-aging	[30]
			Increased collagen synthesis		
		Coffea arabica extract	Higher stability for liposomes /polyhydroxy butyrate	Antioxidant Sunscreen	[31]
	Aspasomes	MAP*	Higher skin permeation and decreased MASI score	Hyperpigmentation	[32]
	Niosomes	Melatonin	Higher skin penetration	Antioxidant Sunscreen	[33]
		EHMC*	Lower skin penetration		
	Transfersomes	RSV*	Higher stability	Antioxidant	[34]
		Myrciaria jaboticaba peel	Higher stability	Antioxidant	[35]

**BP-3: benzophenone-3, EHDP: ethylhexyl dimethyl PABA, EHMC: ethylhexyl methoxycinnamate, BMDBM: methoxydibenzoylmethane, VC-IP: ascorbyl tetraisopalmitate, 4nBR: 4-n-butylresorcinol, RSV: resveratrol, NDAG: naringenin; kaempferol nordihydroguaiaretic acid, MAP: Magnesium ascorbyl phosphate.*

Table 1.
Main outcomes of nanocarriers for cosmetic application.

used lipid nanoparticles in the cosmetic field [36]. SLNs stand out in terms of stability and long-term storage [41].

The advantages of SLNs include (i) possibility of using biodegradable physiological low-toxicity lipids with low toxicity and (ii) ability to form a single-layer film with a hydrophobic character, which provides an occlusive effect on the skin, preventing transepidermal water loss [22]. The occlusion capacity also increases skin elasticity and flexibility, which makes SLN and NLC useful for anti-aging products [22]. In that regard, caffeine-loaded SLN had a faster skin permeation for over 24 hours due to the occlusive effect provided by SLN [21]. The addition of caffeine into skin care products prevents UV-induced photoaging due to its ability to reduce blood flow and antioxidant effects [42].

However, SLNs have shown some drawbacks through the decades: the low entrapment capacity [43], possibility of sudden burst release [44] and the tendency of crystallization of solid lipids [40]. In that context, nanostructured lipid carriers have been developed (NLCs) as a second generation of lipid NPs. NLCs are an advancement to SLNs, bearing a peculiar core composition: a blend of solid lipid and a liquid lipid (e.g. an oil) [40]. The mixture of solid and liquid lipids in NLCs improves stability in comparison to SLNs [3].

Regarding topical application, NLCs provide skin occlusion, skin hydration and sun photoprotection [39]. They are versatile structures allowing the delivery of antioxidants, moisturizers, sunscreens and other bioactive compounds [36]. Similar to SNPs, NLCs improve skin penetration. Coenzyme-Q10 NLC had a higher skin penetration in regards to nanoemulsions and would be an interesting approach to improve antioxidant activity in deeper layers of the skin [23]. Moreover, natural antioxidant compounds encapsulated in NLCs had a prolonged release profile and physicochemical stability for at least 30 days. Therefore, NLCs are effective skin-delivery vehicles for natural antioxidants [24]. In other reports, hydrogel containing vegetable compounds loaded with NLC scavenged free radicals and provided an increase in skin hydration and skin elasticity [25].

The advantage of developing skincare products using both SLNs and NLCs lies in improving the skin absorption of lipid bioactive substances, as nanocarriers increase the cosmetic ingredients solubility. Moreover, their reduced size allows a higher possibility of skin penetration. In addition, once applied to the skin, the nanolipid structures can increase occlusion, which also provides an increase in skin barrier function [3, 36].

Nanoemulsions (NEs) are thermodynamically unstable colloidal systems composed of an oil and aqueous phase and emulsifiers that form lipid-core nanoparticles, with a diameter ranging from 20 to 500 nm [3]. Among the advantages of NE, one can highlight: (i) NEs feature a transparent or translucent appearance; (ii) desired skincare product properties such as low viscosity, great spreadability and pleasant texture; and (iii) enhanced skin hydration [45].

Also, NE can be found as a base for various skincare and cosmetic products, such as deodorants, shampoos, sunscreens, conditioners, and skin and hair serums [46, 47]. NE-based products are among the most popular and commercially available. For instance, Kemira nano-gel is a nanoemulsion-based patented cosmetics system meant to promote skin smoothness [47, 48]. Other commercial cosmetics bearing nanoemulsions are available for skin hydration and skin anti-aging [49].

Besides, NE is acceptable in cosmetics because they are light, transparent and less prone to creaming, sedimentation, flocculation, or coalescence regarding macroemulsions [50]. Further, oil-in-water NEs play an important role in cosmetics as they are fundamental in body lotions, skin creams and sunscreens products. Yet, a recent but fast-growing field of application is wet wipes. It is highly attractive in the growing market for baby care and make-up removal products [51].

4. Cosmetics based on vesicular nanocarriers

Liposomes have been reported as cosmetics carriers since the late 1970s [52] for several skin applications including hydration, anti-aging and sunscreen [6]. They are formed by lipid lamellae and an aqueous core [53] (**Figure 2**), and due to the presence of phospholipids, liposomes have limited stability [5]. Additionally, liposomes usually provide a more superficial skin release, at the stratum corneum level. Hence, other vesicular carriers (**Figure 2**) were developed aiming to obtain more stable formulations and to provide a deeper skin release into the skin [53]. In that regard, transfersomes and niosomes were developed [54]. Transfersomes are vesicular carriers containing phospholipids and an edge activator. Diversely, niosomes are vesicular systems bearing non-ionic surfactants instead of phospholipids [54].

The approach of using plant extracts loaded into vesicular systems [35, 55] and natural-ingredients have been explored as new cosmetics formulations [27, 56, 57]. Moreover, as microbiome is important to skin health and its imbalance is related to disease occurrence, probiotics are also addressed as skin care products [58]. In this sense, probiotics loaded in liposomes reduced the unpleasant odor, an interesting feature to increase adherence to product use, which consequently may contribute to skin health maintenance [27].

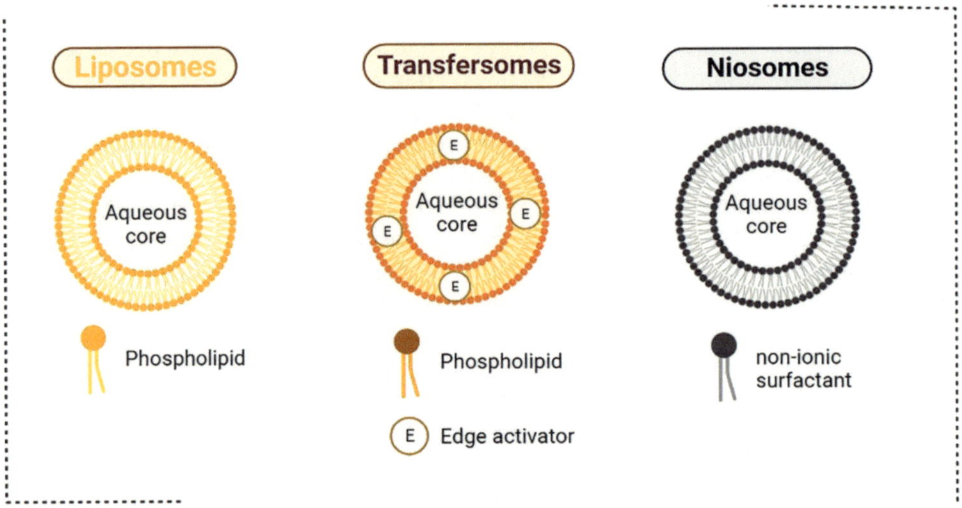

Figure 2.
Vesicular carrriers: Liposomes, transfersomes and niosomes [53, 54].

Regarding skin permeability, deformable liposomes bearing taxifolin or taxifolin tetraoctanoate showed a greater skin permeation compared to non-deformable liposomes [57]. Likewise, aspasomes, lipid vesicles containing magnesium ascorbyl palmitate, had a greater skin permeation. The use of a more stable vitamin C derivative with a more lipophilic nature is an approach to improve skin delivery of vitamin C [32]. Furthermore, octyl methoxycinnamate Pickering emulsions containing melatonin-loaded niosomes provided a deeper penetration for melatonin while also providing a low skin penetration of octyl methoxycinnamate [33]. As melatonin is an antioxidant, its deeper skin penetration may be desirable [33]. On the other hand, octyl methoxycinnamate must remain on the skin surface as it is a sunscreen [59].

Furthermore, liposomes improve formulation stability as well as reduce oxidation of oxidation-susceptible cosmetic ingredients. Cysteamine-loaded liposomes had lower oxidation. As cysteamine has a whitening effect, it may be employed in hyperpigmentation disorders [29]. Similarly, vitamin C and folic acid co-loaded in chitosan-coated liposomes had lower sunlight degradation. Besides, the additional chitosan coating provided an even greater antioxidant ability [26].

As the nanoformulations are designed for skin application, clinical studies are highly recommended to prove their efficacy. In that regard, vesicular carriers improved melasma [28, 32] and skin hydration [56]. Aspasomes [32] and a semisolid bearing resveratrol and n-butylresorcinol encapsulated liposomes [28] reduced melasma in a clinical assessment [28]. Regarding hydration, a cream containing goat milk in liposomes decreased transepidermal water loss. In addition, because there were changes in pH skin, this nanotechnology-based product might be important to prevent barrier function disruption [56].

Cationic carrier is mostly reported as the most suitable one for improving skin penetration [26, 60, 61]. Despite that, anionic liposomes were the most suitable ones for acid ascorbic than cationic liposomes. Anionic ascorbic acid liposomes provided a greater collagen synthesis as well as increased cell uptake, which suggests that the use of anionic carriers for hydrophilic molecules may be a better approach [30].

Therefore, it is essential to conduct a proper physicochemical evaluation prior to nanoformulations development, in order to improve its performance [30].

Recently, the association of polymers and vesicular formulations has been proposed as a strategy to increase carrier stability [31]. In this sense, hydroxyethyl cellulose enriched transfersomes, hyaluronan enriched transfersomes and a hybrid system combining liposomes and poly(3-hydroxybutyrate) had greater physicochemical stability over storage [31, 35]. Accordingly, hybrid systems development may be an interesting approach to obtaining nanocosmetics with higher shelf life [35].

5. Conclusion

Several nanotechnology-based cosmetics have been developed in order to achieve more effective products, with an emphasis on vesicular carriers. Beyond the benefits provided by nanocarriers, nanocosmetics have attracted cosmetic industry interest as they have high market value that may reflect an increased purchase by market consumption.

Therefore, skin delivery of cosmetic ingredients via nanocarriers provides a better alternative to traditional cosmetics by improving skin hydration, skin penetration and stability in addition to their higher versatility. Moreover, herbal–based nanocosmetics may improve patient compliance and have higher biocompatibility, as well as biodegradation potential. Additionally, natural-based nanocosmetics are more environmentally friendly.

Author details

Júlia Scherer Santos[1*], Carolina Sousa Ponciano[2], Thaís Nogueira Barradas[1,2], Mirsiane Pascoal Costa[2] and Guilherme Diniz Tavares[1,2]

1 Faculty of Pharmacy, Department of Pharmaceutical Sciences, Federal University of Juiz de Fora, Juiz de Fora, Brazil

2 Pharmaceutical Sciences Postgraduate Program, Federal University of Juiz de Fora, Brazil

*Address all correspondence to: julia_scherer_santos@hotmail.com

References

[1] Jung MH, Moon JS. A study on the effect of beauty service of the elderly on successful ageing: Focused on mediated effect of self-esteem. Journal of Asian Finance Economics and Business. 2018;**5**(4):213-223. DOI: 10.13106/jafeb.2018.vol5.no4.213

[2] Mcconnel C, Turner L. Medicine, ageing and human longevity. EMBO Reports. 2005;**6**:S59-S62

[3] Salvioni L, Morelli L, Ochoa E, Labra M, Fiandra L, Palugan L, et al. The emerging role of nanotechnology in skincare. Advances in Colloid and Interface Science. 2021;**293**:102437. DOI: 10.1016/j.cis.2021.102437

[4] Dubey SK, Dey A, Singhvi G, Pandey MM, Singh V, Kesharwani P. Emerging trends of nanotechnology in advanced cosmetics. Colloids Surfaces B Biointerfaces. 2022;**214**(March):112440. DOI: 10.1016/j.colsurfb.2022.112440

[5] Santos JS, Barradas TN, Tavares GD. Advances in nanotechnology-based hair care products applied to hair shaft and hair scalp disorders. International Journal of Cosmetic Science. 2022;**44**(3):320-332. DOI: 10.1111/ics.12780

[6] STATNANO. StatNano database [Internet]. 2023. Available from: https://statnano.com/[Accessed: 2023-03-05]

[7] Evangelista M, Mota S, Almeida IF, Pereira MG. Usage patterns and self-esteem of female consumers of antiaging cosmetic products. Cosmetics. 2022;**9**(3):49. DOI: 10.3390/cosmetics9030049

[8] El-Say KM, El-Sawy HS. Polymeric nanoparticles: Promising platform for drug delivery. International Journal of Pharmaceutics. 2017;**528**(1-2):675-691. DOI: 10.1016/j.ijpharm.2017.06.052

[9] Crucho CIC, Barros MT. Polymeric nanoparticles: A study on the preparation variables and characterization methods. Materials Science and Engineering: C. 2017;**80**:771-784. DOI: 10.1016/j.msec.2017.06.004

[10] Olvera-Martínez BI, Cázares-Delgadillo J, Calderilla-Fajardo SB, Villalobos-García R, Ganem-Quintanar A, Quintanar-Guerrero D. Preparation of polymeric nanocapsules containing octyl methoxycinnamate by the emulsifieation-diffusion technique: Penetration across the stratum corneum. Journal of Pharmaceutical Sciences. 2005;**94**(7):1552-1559. DOI: 10.1002/jps.20352

[11] Siqueira NM, Contri RV, Paese K, Beck RCR, Pohlmann AR, Guterres SS. Innovative sunscreen formulation based on benzophenone-3-loaded chitosan-coated polymeric nanocapsules. Skin Pharmacology and Physiology. 2011;**24**(3):166-174. DOI: 10.1159/000323273

[12] Frank LA, Gazzi RP, Pohlmann AR, Guterres SS, Contri RV. Evaluation of an efficient and skin-adherent semisolid sunscreen nanoformulation. Skin Pharmacology and Physiology. 2022;**35**(5):291-298. DOI: 10.1159/000525176

[13] Wu X, Guy RH. Applications of nanoparticles in topical drug delivery and in cosmetics. Journal of Drug Delivery Science and Technology. 2009;**19**(6):371-384. DOI: 10.1016/S1773-2247(09)50080-9

[14] Gilbert E, Roussel L, Serre C, Sandouk R, Salmon D, Kirilov P,

et al. Percutaneous absorption of benzophenone-3 loaded lipid nanoparticles and polymeric nanocapsules: A comparative study. International Journal of Pharmaceutics. 2016;**504**(1-2):48-58. DOI: 10.1016/j.ijpharm.2016.03.018

[15] da Silva NP, Carmo Rapozo Lavinas Pereira E do, Duarte LM, de Oliveira Freitas JC, de Almeida CG, da Silva TP, et al. Improved anti-Cutibacterium acnes activity of tea tree oil-loaded chitosan-poly(ε-caprolactone) core-shell nanocapsules. Colloids Surfaces B Biointerfaces. 2020;196(July):111371. DOI:10.1016/j.colsurfb.2020.111371

[16] Tree-udom T, Wanich wecharungruang SP, Seemork J, Arayachukeat S. Fragrant chitosan nanospheres: Controlled release systems with physical and chemical barriers. Carbohydrate Polymers. 2011;**86**(4):1602-1609. DOI: 10.1016/j.carbpol.2011.06.074

[17] Tsujimoto H, Hara K, Huang C, Yokoyama T, Yamamoto H. Percutaneous absorption study of biodegradable PLGA nanos-spheres via human skin biopsies. Journal of the Society of Powder Technology Japan. 2004;**41**:867-875

[18] Hatem S, ElHoffy N, Elezaby R, Nasr M, Osama A, Elkheshen S. Optimization of the colloidal properties of chitosan nanoparticles encapsulating alpha-arbutin. Archives of Pharmaceutical Sciences Ain Shams University. 2022;**6**(1):17-28. DOI: 10.21608/aps.2022.121464.1081

[19] Oliveira CA De, Peres DDA, Graziola F, Chacra NAB, Araújo GLB De, Flórido AC, et al. Cutaneous biocompatible rutin-loaded gelatin-based nanoparticles increase the SPF of the association of UVA and UVB filters. European Journal of Pharmaceutical Sciences. 2016;**81**:1-9. DOI: 10.1016/j.ejps.2015.09.016

[20] Mejía-Giraldo JC, Scaiano JC, Gallardo-Cabrera C, Puertas-Mejía MA. Photoprotection and photostability of a new lignin-gelatin-baccharis antioquensis-based hybrid biomaterial. Antioxidants. 2021;**10**(12):1904. DOI: 10.3390/antiox10121904

[21] Puglia C, Offerta A, Tirendi GG, Tarico MS, Curreri S, Bonina F, et al. Design of solid lipid nanoparticles for caffeine topical administration. Drug Delivery. 2016;**23**(1):36-40. DOI: 10.3109/10717544.2014.903011

[22] Wissing S. Cosmetic applications for solid lipid nanoparticles (SLN). International Journal of Pharmaceutics. 2003;**254**(1):65-68. DOI: 10.1016/S0378-5173(02)00684-1

[23] Schwarz JC, Baisaeng N, Hoppel M, Löw M, Keck CM, Valenta C. Ultra-small NLC for improved dermal delivery of coenyzme Q10. International Journal of Pharmaceutics. 2013;**447**(1-2):213-217. DOI: 10.1016/j.ijpharm.2013.02.037

[24] Gonçalves C, Ramalho MJ, Silva R, Silva V, Marques-Oliveira R, Silva AC, et al. Lipid nanoparticles containing mixtures of antioxidants to improve skin care and cancer prevention. Pharmaceutics. 2021;**13**(12):2042. DOI: 10.3390/pharmaceutics13122042

[25] Lacatusu I, Istrati D, Bordei N, Popescu M, Seciu AM, Panteli LM, et al. Synergism of plant extract and vegetable oils-based lipid nanocarriers: Emerging trends in development of advanced cosmetic prototype products. Materials Science and Engineering: C. 2020;**108**:110412. DOI: 10.1016/j.msec.2019.110412

[26] Jiao Z, Wang X, Yin Y, Xia J, Mei Y. Preparation and evaluation of a chitosan-coated antioxidant liposome containing vitamin C and folic acid. Journal of

Microencapsulation. 2018;**35**(3):272-280. DOI: 10.1080/02652048.2018.1467509

[27] Htwe MM, Teanpaisan R, Khongkow P, Amnuaikit T. Liposomes of probiotic's lyophilized cell free supernatant; a potential cosmeceutical product. Die Pharmazie. 2019;**74**(8):462-466. DOI: 10.1691/ph.2019.9030

[28] Kwon S, Yang JH, Shin J, Park K, Huh C, Na J. Efficacy of liposome-encapsulated 4-n-butylresorcinol and resveratrol cream in the treatment of melasma. Journal of Cosmetic Dermatology. 2020;**19**(4):891-895. DOI: 10.1111/jocd.13080

[29] Atallah C, Greige-Gerges H, Charcosset C. Development of cysteamine loaded liposomes in liquid and dried forms for improvement of cysteamine stability. International Journal of Pharmaceutics. 2020;**589**(August):119721. DOI: 10.1016/j.ijpharm.2020.119721

[30] Maione-Silva L, de Castro EG, Nascimento TL, Cintra ER, Moreira LC, Cintra BAS, et al. Ascorbic acid encapsulated into negatively charged liposomes exhibits increased skin permeation, retention and enhances collagen synthesis by fibroblasts. Scientific Reports. 2019;**9**(1):522. DOI: 10.1038/s41598-018-36682-9

[31] Pavelkova R, Matouskova P, Hoova J, Porizka J, Marova I. Preparation and characterisation of organic UV filters based on combined PHB/liposomes with natural phenolic compounds. Journal of Biotechnology 2020;**324**(December 2019):100021. DOI:10.1016/j.btecx.2020.100021

[32] Aboul-Einien MH, Kandil SM, Abdou EM, Diab HM, Zaki MSE. Ascorbic acid derivative-loaded modified aspasomes: Formulation, in vitro, ex vivo and clinical evaluation for melasma treatment. Journal of Liposome Research. 2020;**30**(1):54-67. DOI: 10.1080/08982104.2019.1585448

[33] Arslan Azizoglu G, Tuncay Tanriverdi S, Aydin Kose F, Ballar Kirmizibayrak P, Ozer O. Dual-prevention for UV-induced skin damage: Incorporation of melatonin-loaded elastic niosomes into octyl methoxycinnamate Pickering emulsions. AAPS PharmSciTech. 2017;**18**(8):2987-2998. DOI: 10.1208/s12249-017-0786-1

[34] Wu PS, Li YS, Kuo YC, Tsai SJ, Lin CC. Preparation and evaluation of novel transfersomes combined with the natural antioxidant resveratrol. Molecules. 2019;**24**(3):600. DOI: 10.3390/molecules24030600

[35] Castangia I, Manca ML, Allaw M, Hellström J, Granato D, Manconi M. Jabuticaba (Myrciaria Jaboticaba) peel as a sustainable source of anthocyanins and ellagitannins delivered by phospholipid vesicles for alleviating oxidative stress in human keratinocytes. Molecules. 2021;**26**(21):6697. DOI: 10.3390/molecules26216697

[36] Assali M, Zaid AN. Features, applications, and sustainability of lipid nanoparticles in cosmeceuticals. Saudi Pharmaceutical Journal. 2022;**30**(1):53-65. DOI: 10.1016/j.jsps.2021.12.018

[37] Gupta S, Gupta S, Jindal N, Jindal A, Bansal R. Nanocarriers and nanoparticles for skin care and dermatological treatments. Indian Dermatology Online Journal. 2013;**4**(4):267. DOI: 10.4103/2229-5178.120635

[38] Mahant S, Rao R, Nanda S. Nanostructured lipid carriers. In: Design of Nanostructures for Versatile Therapeutic Applications. Amsterdam, Kidlington, Cambridge: Elsevier; 2018. pp. 97-136. DOI: 10.1016/B978-0-12-813667-6.00003-6

[39] Koroleva M, Portnaya I, Mischenko E, Abutbul-Ionita I, Kolik-Shmuel L, Danino D. Solid lipid nanoparticles and nanoemulsions with solid shell: Physical and thermal stability. Journal of Colloid and Interface Science. 2022;**610**:61-69. DOI: 10.1016/j.jcis.2021.12.010

[40] Clares B, Calpena AC, Parra A, Abrego G, Alvarado H, Fangueiro JF, et al. Nanoemulsions (NEs), liposomes (LPs) and solid lipid nanoparticles (SLNs) for retinyl palmitate: Effect on skin permeation. International Journal of Pharmaceutics. 2014;**473**(1-2):591-598. DOI: 10.1016/j.ijpharm.2014.08.001

[41] Ahmad J. Lipid nanoparticles based cosmetics with potential application in alleviating skin disorders. Cosmetics. 2021;**8**(3):84. DOI: 10.3390/cosmetics8030084

[42] Herman A, Herman AP. Caffeine's mechanisms of action and its cosmetic use. Skin Pharmacology and Physiology. 2013;**26**(1):8-14. DOI: 10.1159/000343174

[43] Yoon G, Park JW, Yoon IS. Solid lipid nanoparticles (SLNs) and nanostructured lipid carriers (NLCs): Recent advances in drug delivery. Journal of Pharmaceutical Investigation. 2013;**43**(5):353-362. DOI: 10.1007/s40005-013-0087-y

[44] Ghasemiyeh P, Mohammadi-Samani S. Solid lipid nanoparticles and nanostructured lipid carriers as novel drug delivery systems: Applications, advantages and disadvantages. Research in Pharmaceutical Sciences. 2018;**13**(4):288. DOI: 10.4103/1735-5362.235156

[45] Maali A, Mosavian MTH. Preparation and application of nanoemulsions in the last decade (2000-2010). Journal of Dispersion Science and Technology. 2013;**34**(1):92-105. DOI: 10.1080/01932691.2011.648498

[46] Akram MR, Ahmad M, Abrar A, Sarfraz RM, Mahmood A. Formulation design and development of matrix diffusion controlled transdermal drug delivery of glimepiride. Drug Design, Development and Therapy. 2018;**12**:349-364. DOI: 10.2147/DDDT.S147082

[47] Sonneville-Aubrun O, Simonnet JT, L'Allorest F. Nanoemulsions: A new vehicle for skincare products. Advances in Colloid and Interface Science. 2004;**108-109**(03):145-149. DOI: 10.1016/s0001-8686(03)00146-5

[48] Siddique JA, Husain FM, Singh M. Nanoemulsions: Current trends in skin-care products. In: Setapar SHM, Ahmad A, Jawaid M, editors. Nanotechnology for the Preparation of Cosmetics Using Plant-Based Extracts. Amsterdam, Cambridge: Elsevier; 2022. pp. 49-73. DOI: 10.1016/B978-0-12-822967-5.00005-9

[49] Kaul S, Gulati N, Verma D, Mukherjee S, Nagaich U. Role of nanotechnology in cosmeceuticals: A review of recent advances. Journal of Pharmaceutics. 2018;**2018**:1-19. DOI: 10.1155/2018/3420204

[50] Rocha-Filho PA, Maruno M, Oliveira B, Bernardi DS, Gumiero VC, Pereira TA. Nanoemulsions as a vehicle for drugs and cosmetics. Nanoscience and Technology Open Access. 2013;**1**(1). DOI: 10.15226/2374-8141/1/1/00105

[51] Hada JS. New trends in non-woven wet wipes. International Journal for Modern Trends in Science and Technology. 2020;**06**(9S):89-96. DOI: 10.46501/IJMTST0609S15

[52] Handjani RM, Ribier A, Rondot B, Vanlerberghie G. Dispersions of lamellar phases of non-ionic lipids in cosmetic

products. International Journal of Cosmetic Science. 1979;**1**(5):303-314. DOI: 10.1111/j.1467-2494.1979.tb00224.x

[53] Lai F, Caddeo C, Manca ML, Manconi M, Sinico C, Fadda AM. What's new in the field of phospholipid vesicular nanocarriers for skin drug delivery. International Journal of Pharmaceutics. 2020;**583**:119398. DOI: 10.1016/j.ijpharm.2020.119398

[54] Chen S, Hanning S, Falconer J, Locke M, Wen J. Recent advances in non-ionic surfactant vesicles (niosomes): Fabrication, characterization, pharmaceutical and cosmetic applications. European Journal of Pharmaceutics and Biopharmaceutics. 2019;**144**(August):18-39. DOI: 10.1016/j.ejpb.2019.08.015

[55] Pavaloiu RD, Sha At F, Bubueanu C, Hlevca C, Nechifor G. Design and evaluation of a delivery system based on liposomes for Armoracia rusticana extract. Revista de Chimie. 2019;**70**(7):2347-2349. DOI: 10.37358/RC.19.7.7337

[56] Kocic H, Stankovic M, Tirant M, Lotti T, Arsic I. Favorable effect of creams with skimmed donkey milk encapsulated in nanoliposomes on skin physiology. Dermatologic Therapy. 2020;**33**(4):1-7. DOI: 10.1111/dth.13511

[57] Kim AR, An HJ, Jang ES, Lee JD, Park SN. Preparation, physical characterization, and in vitro skin permeation of deformable liposomes loaded with taxifolin and taxifolin tetraoctanoate. European Journal of Lipid Science and Technology. 2019;**121**(6):1800501. DOI: 10.1002/ejlt.201800501

[58] McLoughlin IJ, Wright EM, Tagg JR, Jain R, Hale JDF. Skin microbiome—The next frontier for probiotic intervention. Probiotics Antimicrob Proteins. 2022;**14**(4):630-647. DOI: 10.1007/s12602-021-09824-1

[59] Mavon A, Miquel C, Lejeune O, Payre B, Moretto P. In vitro percutaneous absorption and in vivo stratum corneum distribution of an organic and a mineral sunscreen. Skin Pharmacology and Physiology. 2007;**20**(1):10-20. DOI: 10.1159/000096167

[60] Elmowafy M. Skin penetration/permeation success determinants of nanocarriers: Pursuit of a perfect formulation. Colloids Surfaces B Biointerfaces. 2021;**203**:111748. DOI: 10.1016/j.colsurfb.2021.111748

[61] Gupta V, Mohapatra S, Mishra H, Farooq U, Kumar K, Ansari MJ, et al. Nanotechnology in cosmetics and cosmeceuticals—A review of latest advancements. Gels. 2022;**8**(3):1-31. DOI: 10.3390/gels8030173

Chapter 3

The Beauty and the Toxic Beast: Use of Comet Assay to Study Antigenotoxicity of Natural Ingredients

Sara Gonçalves and Isabel Gaivão

Abstract

The natural cosmetics market has grown since consumers became conscious of natural-based ingredients. A significant number of cosmetics have noxious and chemically potent substances. Thus, the use of natural and organic cosmetics has become increasingly important. An intense investigation into the benefits fruits and plants can bring to our health is required. A healthy lifestyle can reduce these problems, including the consumption or use of substances that protect the genome through various mechanisms that reduce DNA damage. Genotoxicological studies are essential to know the threats to the genome and health, and antigenotoxicological studies are the answer to minimise the instability of the genome. Natural ingredients such as Almond (*Prunus dulcis*), Elderberry (*Sambucus nigra*), Olives (*Olea europaea*), and Grapes (*Vitis vinifera*) have been shown to possess a variety of biological activities and to hold therapeutic promise. They are the most common ingredients in the Trás-os-Montes region (Portugal). This study aimed to demonstrate, *in vivo*, the genotoxicological effects of Elderberry, Almonds, Olives, and Grapes in *Drosophila melanogaster* using the Comet assay.

Keywords: almonds, antigenotoxicity, comet assay, cosmetics, elderberry, genotoxicity, grapes, natural ingredients, olives

1. Introduction

What is your morning routine? You most likely shower with an invigorating shower gel, some shampoo, and a conditioner with a lovely smell. You also apply a hair mask because you have dry and split hair ends. While the mask sets, you scrub your face with a cleanser; you probably shave if you are a man. Then you apply some facial toner and while it dries, use deodorant under your arms and moisturiser to the rest of your body. Next, you apply moisturiser to your face and some sunscreen. You also brush your teeth with that fresh toothpaste. If you are a lady, you then probably apply some makeup. And do not forget your perfume. According to the Environmental Working Group study, the average man uses five to seven personal care products per day, the average woman uses nine to twelve, and the average teenage girl uses seventeen [1].

A quick look at a cosmetic product's ingredients list shows an enormous amount of noxious and chemically potent substances that impact the environment. A Danish Council THINK Chemicals study from 2020 found that 65 chemicals of concern were found in 39 products [2]. Roughly 69 million individual chemicals and/or chemical combinations are used today [3]. This means consumers are exposed to these chemicals, perhaps daily. The health of our body depends directly on external factors such as cold, heat, humidity, pollution, germs, fungi, bacteria, and the food we eat daily. Our emotions and thoughts also contribute to the maintenance of our health. Healthy habits are associated with life in the open air, good nutrition, and restful sleep. Having beautiful, healthy, silky, and soft skin is the consequence of an excellent functioning organism. Poor digestion and the accumulation of toxins or hormonal imbalances reflect their implications on the skin. To take care of our bodies, we can count on the help of various plants to produce various cosmetics. Environmental elements, air pollution, exposure to solar radiation, and the normal ageing process cause cumulative damage to the building blocks of skin: DNA, collagen, and cell membranes.

The market for natural cosmetics has grown since consumers became conscious of cosmetics with noxious and chemically potent substances and the damage they can cause. People worldwide are striving to make their lifestyle cleaner and safer. As governments begin to act against climate change, growing landfill sites, and the threatening energy crises, we cannot help but consider lifestyle changes and rethink our purchasing habits. Organic has become mainstream, and it is no longer about eating organic food or driving a hydrogen-powered car. The natural and organic market was valued at €76.6 billion at a retail sales price in 2020, and the European cosmetic and personal care cosmetic market is the largest market for cosmetic products worldwide. It is expected to grow annually by 9.24% (2021–2025). There were at least 77 scientific innovation facilities in Europe in 2018 dedicated to research concerning cosmetics and personal care, having spent €2.35 billion in research and development [4, 5].

An intense investigation into the benefits that fruits and plants can bring to our health is needed, as with how to use natural ingredients in cosmetics. A healthy lifestyle can reduce these problems, including the consumption or use of substances that protect the genome through various mechanisms that reduce DNA damage. Genotoxicological studies are essential to know the threats to the genome and health, and antigenotoxicological studies are the answer to minimise the instability of the genome. Natural ingredients such as Elderberry (*Sambucus nigra*), Olives (*Olea europaea*), Almond (*Prunus dulcis*), and Grapes (*Vitis vinifera*) have been shown to have a wide range of biological activities and show promise for therapy. They are the most common ingredients in the Trás-os-Montes region (Portugal). This study aimed to demonstrate, *in vivo*, the genotoxicological effects of Elderberry, Almonds, Olives, and Grapes in *Drosophila melanogaster* using the Comet assay.

2. Skin deep: what is beneath the surface?

The skin has an area of about 2 m^2 and represents about 15% of the body's weight, making it weigh between three and four kilos and is, therefore, the largest and heaviest organ in the human body [6]. The skin covers the entire human organism and is essential for life. Depending on the body region, it has variable appearance, functions, and structure. The skin is a multifunctional organ that provides external coating, thermoregulation, a healthy microbiological environment, and a defence against external aggressions (cold, heat, pressure, pain). It is divided into three layers: epidermis, dermis, and subcutaneous tissue [7], with some of these layers subdivided into further layers, all with specific functions and roles (**Figure 1**).

The epidermis is the most superficial layer of the skin and is in direct contact with the outside. It is stratified and avascularised epithelial tissue and forms the first line of defence against external factors. It subdivides itself into five layers: stratum corneum, stratum lucidum, stratum granulosum, stratum spinosum, and stratum basale (or basal layer). The stratum corneum consists of corneocytes that have migrated up through the skin. After their 30-day life cycle, they start to dry out and flatten, the nuclei disappear, and the corneodesmosomes (a complex protein that holds the corneocytes) begin to break down. At a certain point, the cells start to shed, a process known as desquamation [9]. The dead keratin cells (keratinocytes) reduce the skin's permeability, preventing water loss. Keratinocytes begin life in the base of the epidermis, and as they develop and grow, they move up through the epidermis, pushed up by new cells continuously produced below. They help retain water and form a protective layer that prevents biological, physical, and chemical agents [10]. The epidermis also protects from ultraviolet rays. The sunlight triggers the production of melanin (the brown pigment that defines our skin colour) in the melanocytes located in the

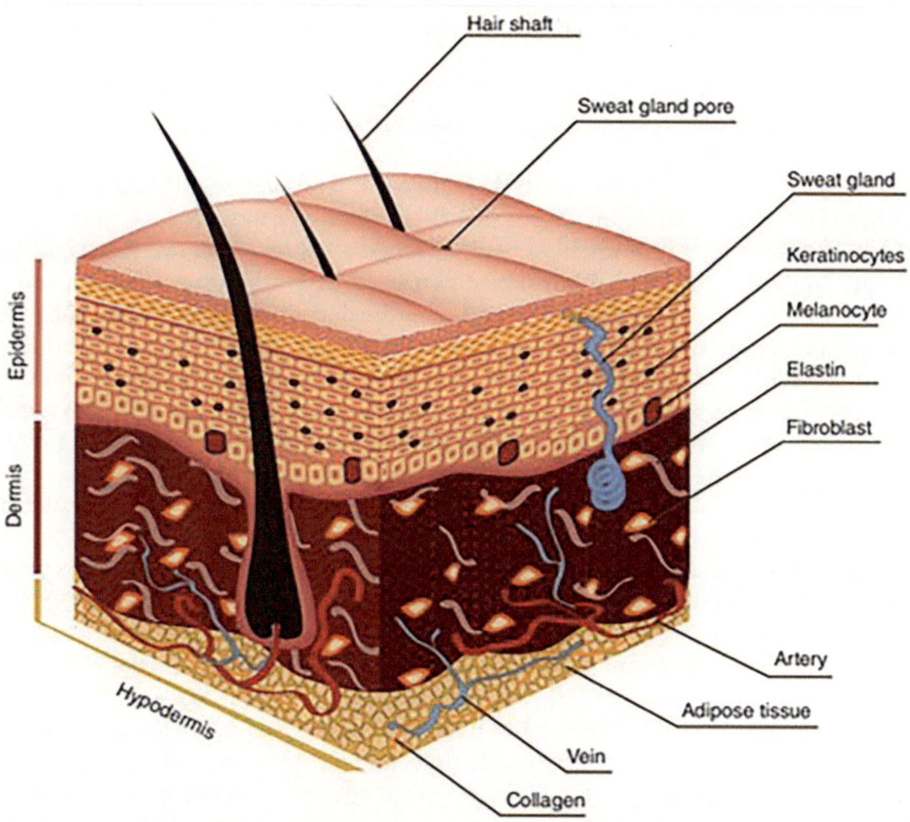

Figure 1.
Representation of the skin. Adapted by Pereira [8].

stratum basale. Melanin acts as a natural sunscreen for the skin and protects it from harmful ultraviolet rays, which is why people get tan when exposed to the sun. But excessive exposure to the sun can disrupt this process, leading to hyperpigmentation. The melanocytes have finger-like projections called dendrites that deliver the melanin into the keratinocytes as they develop and travel up to the stratum corneum [9].

The dermis is a thick, elastic, firm intermediate layer below the epidermis. It subdivides itself into two layers: the reticular layer and the papillary layer [11]. It is composed of irregular connective tissue and has intrinsic collagen fibres, variable elastin fibres, blood vessels, lymphatic vessels, and nerves [10]. It plays an essential role in protecting the body from external and irritating influences, but it also nourishes the upper layers of the skin from the inside. Its thick, firm texture helps to alleviate external pressures. When damage occurs, it contains connecting tissues, such as fibroblasts, that control the production of the extracellular matrix, comprising the structural tissue's collagen and elastin along with water-binding glycoproteins, such as hyaluronic acid [9, 12].

The subcutaneous tissue, which envelops all the muscles except the skin muscles, provides passage for cutaneous nerves, blood, and lymph vessels and plays a role in connecting the dermis and fascia of the muscles [13]. The skin acts like a mesh allowing small molecules to pass freely but screening out larger substances. A substance with a molecular weight or size less than 1000 g/mol can penetrate the skin; with a molecular weight of 400 g/mol, it can enter the cell; with a molecular weight of less than 100 g/mol, it can enter the bloodstream [3]. Smaller materials can easily pass through membranes, enter into DNA, and alter various biochemical functions that are out of our control.

2.1 How and why does the skin's appearance change with age?

Ageing cannot be reversed or even slowed down. However, well-designed cosmetic products can influence the appearance that these changes produce.

The early 20s is when the changes begin. The skin cell turnover starts to slow down gradually. In the 30s, processes that slow down are the natural desquamation, cell turnover, and the rate of collagen production in the dermis. In the 40s, the thickness of the epidermis and dermis decreases, synthesis of collagen and elastin continues to decrease, lipid production slows down, and the glycoproteins begin to decline. The surface becomes more uneven, and fine lines and wrinkles appear as structural integrity declines in the dermis. By the 50s, all processes continue to decline, along with the change in the distribution of subcutaneous fat, the emergence of age spots, and uneven pigmentation. By this stage, all changes lead to apparent effects, such as wrinkles and sagging [9, 14, 15].

The ageing of the skin is primarily associated with the intrinsic genome, as well as with ethnicity (African-American skin is more compacted than Caucasian skin, as well as having a higher intercellular lipid content, which may contribute to more resistance to ageing), anatomical variations (there are significant differences in skin thickness concerning body site, ranging from 0.05 to 0.1 mm on the eyelids to more than 6 mm on the soles of the feet), and hormonal changes in cutaneous tissues (dramatic hormonal changes, particularly thyroid, testosterone, and oestrogen, alter epidermal lipid synthesis) [9, 16, 17]. This type of ageing is known as intrinsic skin ageing.

Alongside this natural ageing process can produce additional ageing effects. These are known as extrinsic ageing factors [17]. These are the ones that each individual can most influence through changes in their lifestyle and the adoption of a good skincare routine.

Skin is affected by ambient conditions such as temperature and humidity. Low temperature stiffens skin and decreases evaporative water loss even with plenty of moisture in the air, as structural proteins and lipids in the skin are critically dependent on temperature for appropriate conformation [18].

Smoking tobacco has been shown to harm the skin. Those who smoke have fewer collagen and elastin fibres in the dermis, which cause the skin to become slack, hardened, and less elastic. Smoking was an independent risk factor for premature wrinkling even when age, sun exposure, and pigmentation were controlled [19].

The effects of sunlight on the skin are profound and are estimated to be responsible for up to 90% of visible skin ageing, known as photoaging [20]. Sunlight includes three different types of radiation: UVC, UVB, and UVA. UVC (100–290 nm) is primarily blocked by the ozone layer and has little effect on the skin. UVB (290–320 nm) penetrates only into the epidermis, being responsible for the erythema associated with a sunburn. UVA penetrates deeper into the skin and may be responsible for most chronic skin damage associated with photoaging [9, 20]. Ultraviolet radiation is a complete carcinogen, as it initiates cancer through DNA mutation and promotes cancer growth through the inflammatory processes inherent in cumulative ultraviolet exposure [21].

Pollution has become a popular topic regarding premature visible skin ageing. It has been shown that cellular senescence and skin ageing are closely regulated and connected and can be induced by air pollution [22].

3. What is a cosmetic?

The word *cosmetic* comes from the Ancient Greek word *kosmos*, their word describing *the art of dress and ornament* [23]. Ornament is used to describe the outer appearance, and letting many believe that cosmetic products are somehow trivial or lacking depth; many other believe that cosmetic product just applies to makeup. However, makeup, often called colour cosmetics or decorative cosmetics, is one class of cosmetic products. **Figure 2** gives an example of the multitude of cosmetic products covered.

According to the definition of the European Regulation, the term cosmetics signifies a product applied to the body to maintain the skin and, thus, the body as a whole, in good condition, to protect it from the environmental influences and ageing processes, to change its appearance, and to enhance the smell of the body [24]. The definition of cosmetics includes shampoos, soaps, toothpaste, colour cosmetics, hair dyes, cleansing and moisturising creams for regular care, styling products, fragrances, and preparation for protection against ultraviolet light (UV light) [25]. Natural, conventional, and organic cosmetics share the same definition but have different specificities. Certified natural and organic ingredients do not need to be present in the formulation of conventional cosmetics [26]. At least one ingredient derived from a natural substance obtained directly from a mineral or a plant must be contained in a natural cosmetic product and must not be produced synthetically. Natural cosmetics may contain percentages of organic ingredients. Nevertheless, natural products are not necessarily organic [27]. An organic cosmetic must contain at least 95% certified organic ingredients. These raw materials are obtained through approved cultivation and extraction. They must be biodegradable and have as natural

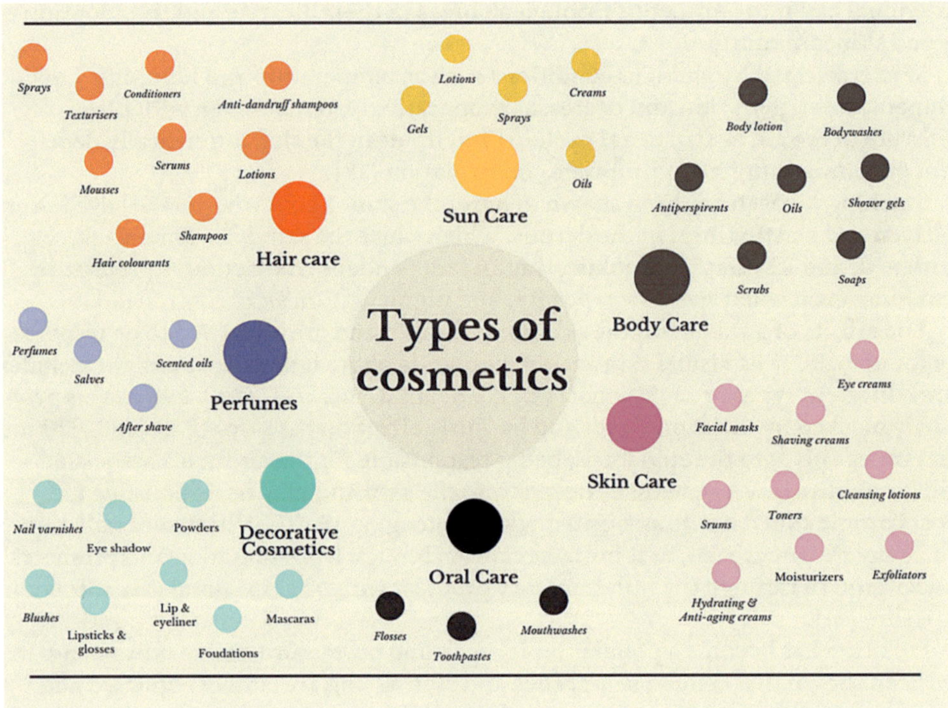

Figure 2.
Diagram showing the range of types of cosmetic products organised and colour-coded in their families.

a chemical nature as possible. The remaining 5% of the formulation may consist of water and natural raw materials from agriculture or non-certified extracting agents approved for organic formulations [25–27].

4. The use of *Drosophila melanogaster* in genotoxicological studies

D. melanogaster was used to evaluate the genotoxic effects caused by natural ingredients exposure. This model organism has become one of the most popular in studies of biological phenomena such as development, neurodegenerative diseases, behaviour, and genotoxicity studies due to its advantages [28–31]. For example, its short life cycle, high prolificacy, and easy handling make it a valuable tool in population studies as it allows the study of several generations in a short period. In addition, their high prolificacy makes statistical analysis easy and reliable [28, 32]. On the other hand, its simple genome, easy manipulation during all phases of development, and the rapid identification of mutants make it a suitable organism in genotoxicity studies, comparative genomics, and evolution [31, 33, 34]. However, the demographic parameters of ageing, such as survival and specific mortality, are highly susceptible even to minor environmental and experimental design variations. Thus, it is necessary to maintain rigorous laboratory practices and carefully control the genetic background to obtain the robust and reliable measurements [35].

The use of *D. melanogaster* as an experimental organism in studies that involve toxicity and genotoxicity has a good level of extrapolation of the results to humans.

Due to the similarities of the detoxification mechanisms, if a compound is antigenotoxic in *D. melanogaster*, it is most likely to be also in humans [36, 37].

5. The comet assay

The alkaline (pH > 13) Comet assay is a sensitive and rapid method for detecting DNA strand breaks in single cells. It is used in genotoxicological studies to determine oxidative DNA damage occurring in various health conditions (in combination with certain bacterial enzymes), to show the protective effects of various nutritional factors in chemopreventive studies, to determine sequence- or gene-specific damage and repair (in combination with fluorescence *in situ* hybridisation) as well as for possible diagnoses [38, 39]. It has the advantages of identifying DNA damage at the single-cell level, sensitivity for detecting low levels of DNA damage using a small number of cells per sample (<10.000), it has a low cost and ease of application, and a short time needed to perform this assay and eukaryote single cell population can be both *in vivo* and *in vitro* [39, 40].

6. Genotoxicity of streptonigrin

Streptonigrin (SN, CAS no. 3930-19-6) is an aminoquinone antitumor antibiotic isolated from cultures of *Streptomyces flocculus* [41]. Due to the potential use of SN in clinical chemotherapy, the study of its genotoxicity is of considerable practical importance. SN inhibits the synthesis of DNA and RNA, causes DNA strand breaks after reduction with NADH, induces unscheduled DNA synthesis and DNA adducts, and inhibits topoisomerase II [42, 43]. This antibiotic causes chromosome damage at the chromosome level and increases the frequency of sister-chromatid exchanges [41]. SN shows its potential to induce a significant level of genotoxicity without the toxic effects (at 20 μM), making it a suitable genotoxic insult for this assay [44, 45].

7. Materials and methods

7.1 Chemicals

Instant Carolina Drosophila Medium Formula 4–24® (hereinafter referred to as Instant Drosophila Medium—IDM) was purchased from Carolina Biological Supply Company, Burlington, USA. Streptonigrin (CAS 3930–19-6) was purchased from Santa Cruz Biotechnology Inc. of Texas, USA. All other chemicals were purchased from Sigma-Aldrich Chemical Company (Madrid, Spain).

7.2 Natural ingredients harvesting and preparation

Elderberry flowers and berries, Olives, Olive Tree Leaves, Grapes, and Almonds were selected in the Trás-os-Montes region, Portugal. This region is bordered by the province of Minho to the west, the Douro region to the south, the Douro River to the east, and Spain to the north. Elderberry is widespread in the north of Portugal, especially in the Varosa Valley, which offers a favourable microclimate for developing this species due to the surrounding mountains [46, 47]. The almond tree is one of the most widespread tree crops in the Trás-os-Montes region, covering an area of 19,206 hectares. The most widespread varieties are *Parada*, *Casanova*, *Verdeal*, and *Pegarinhos* [48, 49]. Portugal has a wine-growing area of 1/4 to 1/5 of the surface of the important wine-growing countries of Europe. Of the 343 grape varieties listed, about 230 are considered native to Portugal or the Iberian Peninsula, reflecting the extensive and unique Portuguese viticultural genetics [50]. Trás-os-Montes is the second most important Portuguese olive-growing area, currently accounting for between 12 and 15% of national olive oil production. The most important varieties are *Cobrançosa*, *Madural,* and *Verdeal* [51, 52]. In the Trás-os-Montes region, 40 native varieties are grown [50]. Therefore, natural ingredients are easy to obtain in this area. It is also the region with the most organic farmers, and the climatic, topographical, and pedological differences predestine this region for agricultural diversity [53].

Almonds (variety *Pegarinhos*), Red Grapes (variety *Touriga Nacional*), Olive Tree Leaves, and Olives (variety *Cobrançosa*) were obtained from organic farmers in October, September, and December 2021. Elderberry flowers and berries were harvested at Vila Verde, Alijó, Portugal (41°21′44.2"N, 7°33′01.9"W) in May 2022 for the flowers and August 2022 for the berries. Before the experiments, natural ingredients were ground with a coffee mill, obtaining particles <2 mm. After that, they were collected in an airless plastic bag and frozen at -18°C until further analysis.

7.3 Drosophila stock

D. melanogaster Oregon K (Ok) strain was chosen since it has a low antioxidant enzymatic activity and is, therefore, more sensitive for this study [44]. For the genotoxicity assay, crossings were made to obtain heterozygous offspring (w/w^+). Flies were kept in an incubator at 24°C and were anesthetised by etherisation when necessary.

7.4 Genotoxicity evaluation

The evaluation of the genotoxic/antigenotoxic effects of these ingredients was carried out through the Comet assay.

7.4.1 Comet assay

The assay was performed based on the described method [54]. Based on the results obtained in a previous study, for each natural ingredient quantity tested (Elderberry: 5 g; Elderberry flower: 10 g; Olive: 10 g; Olive leaf: 1 g; Grape pulp: 10 g; Almond: 10 g; Almond shell: 1 g; each in a final volume of 100 mL of medium) five instar larvae were isolated and placed in a Petri dish to isolate the neuroblasts. In the end, eight microtubes with five brain ganglia each, immersed in Ringer solution, were obtained. The next step was neuroblast maceration, followed by centrifugation, supernatant removal, and the addition of 140 μL of low melting point (LMP) agarose. Two 70 μL of this solution was then placed in a slide precoated with 10% agarose normal melting point (NMP). Each drop was covered with a coverslip, spreading the solution. This process was repeated for each quantity tested. The slides were stored at 4°C for 5 min, and after agarose solidification, the coverslips were removed. The slides were then immersed in cold fresh lysis solution (2.5 M NaCl, 0.1 M ethylenediaminetetraacetic acid disodium salt (EDTA), 0.01 M Tris base, 1% Triton X-100, pH 10). For the preparation of the lysis solution, the correct amount of all compounds was dissolved in distilled water (less than the final volume), and pH was set to 10 with 10 M NaOH solution) for 1 h at 4°C, after which they were placed in the electrophoresis chamber with their frozen ends to the cathode and without empty spaces among them. The electrophoresis chamber was filled with cold denaturing and electrophoresis buffer until the slides were covered and stored for 20 min. The next step was the electrophoresis in the dark, where a current of 300 mA and a voltage of 25 V (corresponding a 0,8 V/cm) were applied for 20 min. In the end, the slides were washed in PBS for 10 min at 4°C, then in distilled water for 10 min at 4°C, and left to air-dry. The slides were stained with 40 μL of DAPI (1 μg/mL in water) to each gel and covered with a coverslip. The slide analysis was performed using a fluorescence microscope, and the % of tail DNA and the tail length were scored. For this, 50 cells per gel were observed, and each cell was classified from 0 (no tail) to 4 (almost all DNA in the tail) based on the intensity of its tail. The final score (expressed as "arbitrary units" in a range of 0—400) was obtained by multiplying the mean percentage of nucleoids in each class by the appropriate factor according to this formula:

$$\begin{aligned} \textit{Genetic Damage Indicator}\,(GDI) = {} & [(\%\textit{nucleoid class}\,0)\,x\,0)] + [(\%\textit{nucleoid class}\,1)\,x\,1)] \\ & + [(\%\textit{nucleoid class}\,2)\,x\,2)] + [(\%\textit{nucleoid class}\,3)\,x\,3)] \\ & + [(\%\textit{nucleoid class}\,4)\,x\,4)] \end{aligned} \quad (1)$$

Data were analysed using the software IBM SPSS Statistics (Statistical Package for the Social Sciences, Chicago, IL, USA), version 20. An analysis of variance (ANOVA) was performed, followed by a Tukey test. Differences were considered statistically significant if $p < 0.05$.

The same procedure was used for the streptonigrin challenge. Considering the final medium volume, streptonigrin was added to the IDM, and dissolved in PBS to attain the final concentration of 20 μM. This concentration was selected according to the literature [44]. A schematic representation of the experimental design can be observed in **Figure 3**.

8. Results

The streptonigrin-challenged group exhibited overall increased DNA damage in all ingredients assessed. Flies fed with C and challenged with streptonigrin presented the highest levels of DNA damage, while flies fed with Eb showed the lowest levels in both unchallenged and streptonigrin-challenged groups (**Figure 4**). Regarding the unchallenged group, there are no statically significant differences. As for the SN-challenged group, there are differences between SN with OTL, GP, and Eb. Statically significant differences were observed between Eb with OP, AS, and A.

9. Discussion

In this Comet assay, flies fed with Elderberry showed the lowest levels in both unchallenged and streptonigrin-challenged groups.

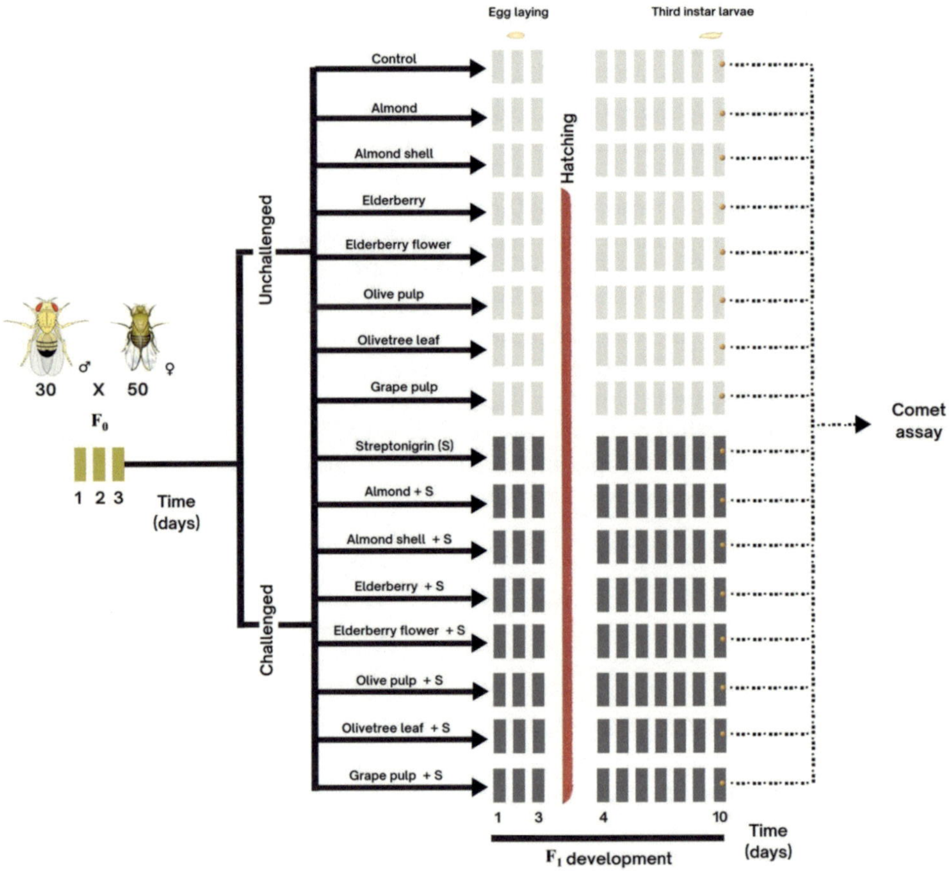

Figure 3.
Schematic representation of the experimental design, elucidating the division of D. melanogaster individuals into two major groups: One unchallenged (light grey time scale) and another streptonigrin-challenged (dark grey challenged) with streptonigrin (SN). The comet assay was performed on larvae in the 3rd instar stage, approximately 10 days after the egg laying.

The identification of ingredients with antigenotoxic effects is one of the most promising areas of research in recent years, as they could protect against DNA damage and its consequences. Studies have shown that antigenotoxic properties are associated with anti-ageing properties [55, 56], and these properties are important in reversing genotoxic effects. Our cells are attacked not only through our skin but also through all the elements we are exposed to. Cosmetics with antigenotoxic properties are capable of neutralising those toxic effects.

10. Conclusions

Beauty is skin deep. The human skin is a powerful organ that seems to be constantly hungry for anything that touches its surface. Oxygen, nitrogen, carbon dioxide, and toxic pollutants enter our skin through three doors: the sweat ducts, hair follicles, and sebaceous glands, or directly through the stratum corneum. This ability

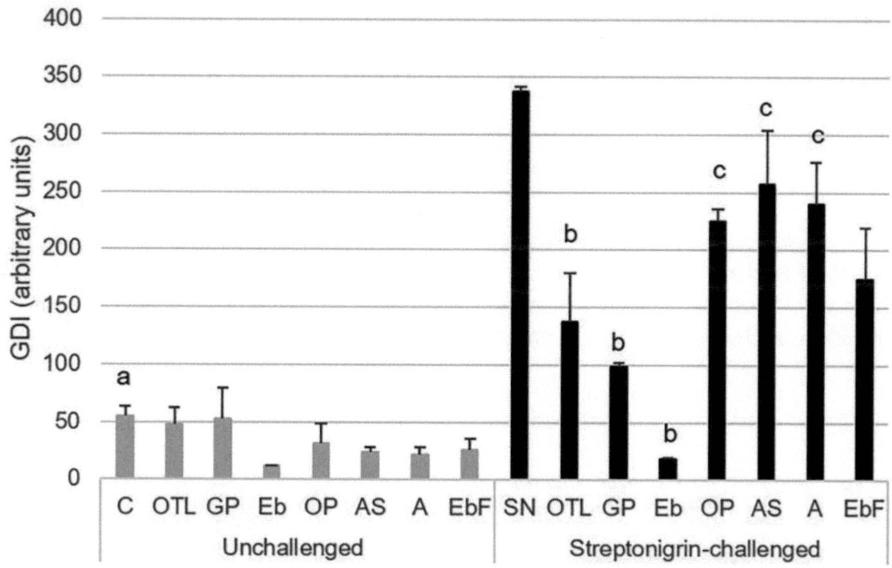

Figure 4.
Mean values of DNA damage (GDI) in drosophila neuroblasts measured by the in vivo comet assay in unchallenged and streptonigrin-challenged groups. Tested groups are identified by abbreviations identifying the ingredient (Eb: Elderberry; OTL: Olive tree leaf; a: Almond; AS: Almond Shell; EbF: Elderberry flower; GP: Grape pulp; OP: Olive pulp). Grey bars correspond to unchallenged groups, and black bars correspond to streptonigrin-challenged groups. Values are mean ± SEM (n = 2). A: Statically significant relative to C; b: Statically significant relative to SN; c: Statically significant relative to BS with treatment.

of the skin to absorb chemical substances so that they can spread throughout the body is often used in medicine. Our skin can absorb up to 64% of substances applied to its surface [57]. Unfortunately, along with water, vitamins, minerals, and oxygen, the skin soaks up potentially carcinogenic ingredients that increase our risk of cancer or other diseases. Natural ingredients hold beneficial properties that need to be studied to be used in the cosmetic industry, benefiting humans. The introduction of antigenotoxic ingredients in cosmetics has the great advantage that, in addition to not being genotoxic per se, and they also neutralise genotoxicity induced by other environmental factors, such as pollution, drugs.

All tested ingredients presented antigenotoxicological properties, with Elderberry having the best results. Such potential cannot be ignored, and further investigation into how to incorporate this ingredient in a cosmetic should occur. Only two studies have been developed since 2012, showing that Elderberry has no mutagenic effect and a high *in vitro* activity [58]. However, none of them focused on cosmetic ingredients.

Additional investigation can be carried out, namely studying antigenotoxicological properties in human lymphocytes. Human lymphocytes are used as surrogate tissue, as they are easily obtained, are available in large numbers, do not require cell culture, are diploids, and are almost all in the same cell cycle phase. Although the Comet assay is well accepted among the scientific community, there are issues regarding standardisation among laboratories [59]. Therefore, new methods for DNA damage assessment would be beneficial to improve research on DNA damage repair and antigenotoxicity.

Acknowledgements

This work was supported by the project UIDB/00772/2020, funded by the Fundação para a Ciência e Tecnologia (FCT).

The authors would like to thank Paula Santenico and Dinis Gonçalves, organic farmers, for providing the ingredients used in this research.

Conflict of interest

The authors declare no conflict of interest.

Author details

Sara Gonçalves[1,2]* and Isabel Gaivão[1,2]

1 Department of Genetics and Biotechnology and CECAV, University of Trás-os-Montes and Alto Douro, Vila Real, Portugal

2 Associate Laboratory for Animal and Veterinary Sciences (AL4AnimalS), Portugal

*Address all correspondence to: sgoncalves@utad.pt

References

[1] Exposures add up – Survey results, Environmental Working Group. Available from: https://www.ewg.org/news-insights/news/2004/12/exposures-add-survey-results. [acedido 3 de outubro de 2022]

[2] Jørgensen C. Cosmetics Worldwide – Same Contents? A Comparative Study. Copenhagen, Denmark: The Danish Consumer Council THINK Chemicals; 2020

[3] Michalun MV, Dinardo EJC. Skin Care and Cosmetic Ingredients Dictionary. 4th ed. NY, USA: Milady - Cenage Learning; 2015

[4] Statista, Global market value for natural/organic cosmetics and personal care in 2018-2027, Statista, 2018. Available from: https://www.statista.com/statistics/673641/global-market-value-for-natural-cosmetics/. [acedido 12 de julho de 2021]

[5] Cosmetics Europe, Cosmetics Industry, Cosmetics Europe - The Personal Care Association. 2021. Available from: https://cosmeticseurope.eu/cosmetics-industry. [acedido 12 de julho de 2021]

[6] M. P. V. Kede, E. O. Sabatovich, Dermatologia estética, Dermatologia estética. 2004. pp. 771-771. Available from: https://pesquisa.bvsalud.org/portal/resource/pt/lil-442728. [Acedido: 18 de janeiro de 2021]

[7] Jean K. Anatomy, histology and immunohistochemistry of normal human skin. European Journal of Dermatology. 2002;**12**(4):390-399; quiz 400

[8] Pereira RF, Barrias CC, Granja PL, Bartolo EPJ. Advanced biofabrication strategies for skin regeneration and repair. Nanomedicine. 2013;**8**(4):603-621. DOI: 10.2217/nnm.13.50

[9] Barton S. Discovering Cosmetic Science. United Kingdom: Royal Society of Chemistry; 2021

[10] Seeley R, VanPutte C, Regan J, Russo EA. Seeley's Anatomy & Physiology. 0th ed. United States of America: McGraw Hill; 2014

[11] Graham HK, Eckersley A, Ozols M, Mellody KT, Sherratt EMJ. In: Limbert G, editor. Human Skin: Composition, Structure and Visualisation Methods, Em Skin Biophysics. Vol. 22. Cham: Springer International Publishing; 2019. pp. 1-18. DOI: 10.1007/978-3-030-13279-8_1

[12] Gonçalves S. Use of enzymes in cosmetics: Proposed enzymatic peel procedure. CosmEthically Active. 2021;**1**:29-35

[13] Ishida T et al. Anatomical structure of the subcutaneous tissue on the anterior surface of human thigh. Okajimas Folia Anatomica Japonica. 2015;**92**(1):1-6. DOI: 10.2535/ofaj.92.1

[14] Ganceviciene R, Liakou AI, Theodoridis A, Makrantonaki E, Zouboulis ECC. Skin anti-aging strategies. Dermato-Endocrinology. 2012;**4**(3):308-319. DOI: 10.4161/derm.22804

[15] Farage MA, Miller KW, Elsner P, Maibach EHI. Characteristics of the aging skin. Advances in Wound Care. 2013;**2**(1):5-10. DOI: 10.1089/wound.2011.0356

[16] Graf J. Anti-Aging Skin Care Ingredient Technologies, Em Cosmetic

Dermatology. Berlin, Heidelberg: Springer; 2005. pp. 17-28

[17] Farage MA, Miller KW, Elsner P, Maibach EHI. Intrinsic and extrinsic factors in skin ageing: A review. International Journal of Cosmetic Science. 2008;**30**(2):87-95. DOI: 10.1111/j.1468-2494.2007.00415.x

[18] McCallion R, Po EALW. Dry and photo–aged skin: And manifestations management. Journal of Clinical Pharmacy and Therapeutics. 1993;**18**(1):15-32. DOI: 10.1111/j.1365-2710.1993.tb00562.x

[19] Leow Y-H, Maibach EHI. Cigarette smoking, cutaneous vasculature and tissue oxygen: An overview. Skin Research and Technology. 1998;**4**(1):1-8. DOI: 10.1111/j.1600-0846.1998.tb00077.x

[20] Gilchrest BA. A review of skin ageing and its medical therapy. British Journal of Dermatology. 1996;**135**(6):867-875. DOI: 10.1046/j.1365-2133.1996.d01-1088.x

[21] Green BG, Bluth EJ. Measuring the chemosensory irritability of human skin, journal of toxicology. Cutaneous and Ocular Toxicology. 1995;**14**(1):23-48. DOI: 10.3109/15569529509057664

[22] Martic I, Jansen-Dürr P, Cavinato EM. Effects of air pollution on cellular senescence and skin aging. Cell. 2022;**11**(14):2220. DOI: 10.3390/cells11142220

[23] Cosmetic - Wiktionary. Available from: https://en.wiktionary.org/wiki/cosmetic (acedido 24 de outubro de 2022)

[24] Singh SK. Handbook on Cosmetics (Processes, Formulae with Testing Methods). Delhi, India: Asia Pacific Business Press Inc.; 2010

[25] EUR-Lex - 31976L0768 - EN, Official Journal L 262, 27/09/1976 P. 0169 - 0200; Greek special edition: Chapter 13, Volume 4, P. 0145; Spanish special edition: Chapter 15, Volume 1, P. 0206; Portuguese special edition, Chapter 15 Volume 1, P. 0206; Finnish special edition: Chapter 13, Volume 5, P. 0198; Swedish special edition: Chapter 13, Volume 5, P. 0198;1997

[26] Romero V, Khury E, Aiello LM, Leonardi EGR. Differences between organic and natural cosmetics: Clarifying literature for prescribers. Surgical and Cosmetics Dermatology. 2018;**10**(3):188-193. DOI: 10.5935/scd1984-8773.20181031087

[27] Fonseca-Santos B, Corrêa M, Chorilli EM. Sustainability, natural and organic cosmetics: Consumer, products, efficacy, toxicological and regulatory considerations. Brazilian Journal of Pharmaceutical Science. 2015;**51**:17-26. DOI: 10.1590/S1984-82502015000100002

[28] Stocker H, Gallant EP. In: Dahmann C, editor. Getting Started: An Overview on Raising and Handling Drosophila, Em Drosophila. Vol. 420. Totowa, NJ: Humana Press; 2008. pp. 27-44. DOI: 10.1007/978-1-59745-583-1_2

[29] Botella JA, Bayersdorfer F, Gmeiner F, Schneuwly ES. Modelling Parkinson's disease in drosophila. NeuroMolecular Medicine. 2009;**11**(4):268-280. DOI: 10.1007/s12017-009-8098-6

[30] Moloney A, Sattelle DB, Lomas DA, Crowther EDC. Alzheimer's disease: Insights from Drosophila melanogaster models. Trends in Biochemical Sciences. 2010;**35**(4):228-235. DOI: 10.1016/j.tibs.2009.11.004

[31] Markow T, O'Grady EP. Drosophila: A Guide to Species Identification and Use. Vol. 259. Massachusetts,

EUA: Elsevier; 2005. DOI: 10.1017/CBO9781107415324.004

[32] Zuk M, Garcia-Gonzalez F, Herberstein ME, Simmons ELW. Model systems, taxonomic bias, and sexual selection: Beyond *drosophila*. Annual Review of Entomology. 2014;**59**(1):321-338. DOI: 10.1146/annurev-ento-011613-162014

[33] Miller C. Drosophila melanogaster, Animal Diversity Web. Available from: https://animaldiversity.org/accounts/Drosophila_melanogaster/. [acedido 6 de junho de 2022]

[34] Drosophila melanogaster-Life Cycle. Available from: https://studyres.com/doc/1271098/drosophila-melanogaster-life-cycle. [acedido 6 de junho de 2022]

[35] Linford NJ, Bilgir C, Ro J, Pletcher ESD. Measurement of lifespan in Drosophila melanogaster. JoVE. 2013;**71**:50068. DOI: 10.3791/50068

[36] Silva GND, de Souza AP, Conter FU, Lehmann M, Dihl ERR, Estudo in silico da interação do δ9-tetrahidrocanabinol com processos biológicos em drosophila melanogaster E Homo sapiens, Revista de Iniciação Científica da ULBRA, no. 18, Art. 2020. Available from: http://www.periodicos.ulbra.br/index.php/ic/article/view/6144 [Acedido: 4 de novembro de 2022]

[37] Graf U, The Actual SMART (Somatic Mutation and Recombination Test) in D. Melanogaster, Contam. Vol. 10, No. 1. London, UK: Ambient; 1994. pp. 5-7

[38] Smith KEC. The control of hydrophobic compound exposure in In vitro tests for Genotoxicity, em Genotoxicity and DNA Repair, L. M. Sierra, E. I. Gaivão, Eds. New York, NY: Springer New York, 2014. pp. 59-72. DOI: 10.1007/978-1-4939-1068-7_4

[39] Liao W, McNutt MA, Zhu EW-G. The comet assay: A sensitive method for detecting DNA damage in individual cells. Methods. 2009;**48**(1):46-53. DOI: 10.1016/j.ymeth.2009.02.016

[40] Kumaravel TS, Vilhar B, Faux SP, Jha EAN. Comet assay measurements: A perspective. Cell Biology and Toxicology. 2009;**25**(1):53-64. DOI: 10.1007/s10565-007-9043-9

[41] Bolzán AD, Bianchi EMS. Genotoxicity of streptonigrin: A review. Mutation Research/Reviews in Mutation Research. 2001;**488**(1):25-37. DOI: 10.1016/S1383-5742(00)00062-4

[42] Cohen MM, Shaw MW, Craig EAP. The effects of streptonigrin on cultured human leukocytes. Proceedings. National Academy of Sciences. United States of America. 1963;**50**(1):16-24. DOI: 10.1073/pnas.50.1.16

[43] Kim SA, Kwon Y, Kim JH, Muller MT, Chung EIK. Induction of topoisomerase II-mediated DNA cleavage by a protoberberine alkaloid, Berberrubine. Biochemistry. 1998;**37**(46):16316-16324. DOI: 10.1021/bi9810961

[44] Gaivão I, Comendador EMA. The w/w+ somatic mutation and recombination test (SMART) of Drosophila melanogaster for detecting reactive oxygen species: Characterization of 6 strains. Mutation Research/Environmental Mutagenesis and Related Subjects. 1996;**360**(2):145-151. DOI: 10.1016/0165-1161(96)00003-9

[45] Gaivão I, Sierra LM, Comendador EMA. The w/w+ SMART assay of Drosophila melanogaster detects the genotoxic effects of reactive oxygen species inducing compounds. Mutation Research/Genetic Toxicology and Environmental Mutagenesis.

1999;**440**(2):139-145. DOI: 10.1016/S1383-5718(99)00020-0

[46] Braga FG et al. Variation of the anthocyanin content in Sambucus nigra L. populations growing in Portugal. Journal of Herbs Spices & Medicinal Plants. 2002;**9**(4):289-295. DOI: 10.1300/J044v09n04_05

[47] Trindade C, Valdiviesso T, de Oliveira EPB. Caraterização morfológica das inflorescências de variedades de Sambucus nigra L, Monografia. Vila Real, Portugal: Universidade de Trás-os-Montes e Alto Douros; 2019

[48] Centro Nacional de Competências dos Frutos Secos, Amêndoa. Estudo de produção e comercialização nas Terras de Trás-os-Montes. Bragança, Portugal: CNCFS; 2020

[49] Cordeiro V, Monteiro EA. Almond growing in trás-os-montes region (Portugal). Acta Horticulturae. 2002;**591**:161-165. DOI: 10.17660/ActaHortic.2002.591.22

[50] Sousa M, Pererira C, Guerra J, Abade EE. Caracterização de Castas Cultivadas na Região Vitivinícola de Trás-os-Montes, Sub regiões de Chaves, Planalto Mirandês e Valpaços, Coleção uma agricultura com norte. Mirandela, Portugal: Ministério da Agricultura, do Desenvolvimento Rural e das Pescas; 2007. p. 40

[51] Portal do INE. 2013. Available from: https://www.ine.pt/xportal/xmain?xpid=INE&xpgid=ine_indicadores&indOcorrCod=0000708&xlang=pt&contexto=bd&selTab=tab2 (acedido 16 de novembro de 2020)

[52] de Figueiredo T, Almeida A, Araújo EJ. Edaphic characteristics of olive-tree areas in the trá-os-montes region (Portugal): A map-based approach. Acta Horticulturae. 2002;**586**:151-154. DOI: 10.17660/ActaHortic.2002.586.24

[53] Gonçalves S, Gaivão EI. Assessment of ANTIGENOTOXIC properties IN natural ingredients common IN the TRÁS-OS-MONTES region: A PHD project. Vol. 1. Berlim, Germany: ScienceOpen; 2021. p. 1. DOI: 10.13140/RG.2.2.11406.28482

[54] María SL, Isabel G, editors. Genotoxicity and DNA Repair. New York, NY: Springer; 2014. DOI: 10.1007/978-1-4939-1068-7

[55] Izquierdo-Vega J et al. Evidence of some natural products with Antigenotoxic effects. Part 1: Fruits and polysaccharides. Nutrients. 2017;**9**(2):102. DOI: 10.3390/nu9020102

[56] Boran R. Investigations of anti-aging potential of Hypericum origanifolium Willd. For skincare formulations. Industrial Crops and Products. 2018;**118**:290-295. DOI: 10.1016/j.indcrop.2018.03.058

[57] Brown HS, Bishop DR, Rowan ECA. The role of skin absorption as a route of exposure for volatile organic compounds (VOCs) in drinking water. American Journal of Public Health. 2011;**74**(5):479-484. DOI: 10.2105/AJPH.74.5.479

[58] Gonçalves S, Gaivão EI. Natural ingredients common in the Trás-os-Montes region (Portugal) for use in the cosmetic industry: A review about chemical composition and Antigenotoxic properties. Molecules. 2021;**26**(17):5255. DOI: 10.3390/molecules26175255

[59] Møller P, Möller L, Godschalk RWL, Jones EGDD. Assessment and reduction of comet assay variation in relation to DNA damage: Studies from the European comet assay validation group. Mutagenesis. 2010;**25**(2):109-111. DOI: 10.1093/mutage/gep067

Chapter 4

Enhancing Skin Cicatrization with Natural Sources – The Role of Polyunsaturated Fatty Acids (PUFAs) and Beeswax

Irina Saretzky and Marta Cassini

Abstract

Objectives: To evaluate the effectiveness and tolerance in patients with RDEB of an ointment with the following active ingredients: petroleum jelly (30%), cod liver oil (10%), beeswax, sunflower oil, BHT, sweet almond oil. The formulation does not contain antibiotics or corticosteroids. Methods: A prospective study was carried out on a cohort of ten patients diagnosed with RDEB confirmed by molecular biology that are followed at the Centre for Research in Genodermatosis and EB. Given the seriousness of the pathology and the difficulties in obtaining good results with the therapeutic resources available, a one-year pilot test was performed with the use of an off-label drug. The therapeutic effectiveness was evaluated as well as the tolerance and safety of the ointment. Patients with infected injuries, anaemic with hemoglobin levels lower than 8 g/dl or albumin levels lower than 3 g/dl were excluded from the study. Before starting the treatment, a complete blood test was carried out. Lesions were evaluated by a grading system considering the compromise, depth and edges of the injury. Photographs of the injuries were taken at the beginning and at weekly controls for the first month and later monthly. The ointment was applied after bathing on sterilised dressings, followed by the placement of secondary dressings that allowed the oxygenation of the lesions. Both applications were changed every 12 or 24 hours. Results: The drug was evaluated for the period of one-year in each patient. No adverse effects were observed. It was well tolerated and blood tests results were not affected. After 48 hours of initiating the treatment the patients reported: absence of pain, increase of average sleep, decrease in the use of analgesic medication, faster healing of blisters in the skin, diminishing the affected areas and significant improvement in life quality. Conclusions: Even though the number of patients in the study was small, the results obtained allow us to conclude that this medication may be beneficial to alleviate the signs and symptoms of patients with RDEB. Subsequent studies on a greater number of patients will allow the confirmation of the seen couraging results.

Keywords: polyunsaturated fatty acids, beeswax, natural source, wound healing, inflammation

1. Introduction

Wound healing has become a field of interest for biotechnological developments which have led in the last 10 years to an intense search for new findings in terms of synthetic compounds with healing properties, drug delivery systems and synthetic fibers and polymers, biomaterials, and nanoparticles [1, 2].

Skin cicatrization is a living model of the regenerative tissue process. Many comorbidities for instance bioburden and antibiotic resistance can appear during cicatrization and because of antibiotic abuse, and there is an increasing percentage of antibiotic resistance in developing countries. Another issue is wound bed inflammation which can be a chronic condition and establish a lack of healing timeline.

Despite technological advances and discoveries in the wound healing field, there is a turning back point in which doctors and researchers all over the world are paying attention to natural source agents with skin restorative capacity.

Beeswax for instance can provide antimicrobial peptides, which can prevent local infection at wound sites [3–6]. Polyunsaturated fatty acids (PUFA's) from cod liver oil, sunflower oil, and sweet almond oil provide a natural source of omega 3–6 with a local anti-inflammatory action at the wound bed [7, 8]. Recent case reports had reported the benefit of a combination of both PUFAs and beeswax for the topical treatment of sacral sores, burn wounds, and severe skin conditions like epidermolysis bullosa [9–12].

Natural resources with a medical scope and careful production process should lead to an old and new pathway for skin regeneration. Medicine and Technology should work together to find a balance to use natural molecules capable of reducing local inflammation, and preventing and controlling the local infection.

It is very common for topical wound healing products to provoke contact dermatitis reactions that lead to discontinuing the product.

In this chapter, the authors will make an overview throughout the cicatrization process of the human skin, including frequent causes that alter wound healing like biofilm and local infection, pressure sores and hypoxia, burns, and genetic disorders which share great denuded skin areas. Authors will also describe their clinical experience in both adult and pediatric patients with chronic wounds treated with an ointment rich in PUFAs, ceramides, and antimicrobial peptides, Curefini™.

2. Cicatrization process. Stages

Wound healing is a well-organized process that can be separated into four sequenced and overlapped stages: hemostasis, inflammation, proliferation, and maturation. These can normally occur in acute wounds like in a knife cut for example.

Hemostasis is a coordinated mechanism of preservation to stop hemorrhage and avoid depleting the blood volume. Instantly, blood vessels constrict to restrict blood flow. Next, platelets aggregate to plug in the damage area in the wall of the blood vessel (platelet activation) followed by a coagulation cascade in which participate multiple factors that contribute to definitive fibrin clot formation. The fibrin clot is a provisory matrix that scaffolds the reconstruction of the dermis matrix of the human skin [13].

Even do inflammation has been presented as the second stage of the wound healing process, the presence of the immune system occurs from the very beginning of the injury throughout the granulation phase also. The initial infiltrate is composed of neutrophil cells to remove cellular debris and bacteria and decontaminate the wound bed. Then, neutrophils are replaced after by a mononuclear infiltrate. Celso's tetrade is often present in this stage: swelling, heat, pain, and redness. Inflammation is a natural part of the wound healing process and is only problematic if prolonged or excessive.

The proliferative or granulation phase is ongoing all the time in the background. By the first week, the fibroblasts have started to deposit new collagen and glycosaminoglycans. These proteoglycans form the core of the wound and help stabilize the wound. Then, reepithelialization starts to occur with the migration of cells from the wound periphery and adjacent edges. The granulation phase not only implies the reconstruction of a new collagen net at the dermis but also the presence of new vessels to provide oxygen and nutrients [14]. In healthy stages of wound healing, granulation tissue is pink or red. When occlusive or semiocclusive dressings are applied within 48 hours after injury, correct tissue humidity and hydration are maintained to optimize epithelialization. Initially, only a thin superficial layer of epithelial cells is laid down, but a thicker and more durable layer of cells will bridge the wound over time. Fibroblasts suffer a phenotype change to myofibroblasts. Myofibroblast-acquired contractile cytoskeleton allows them to cause the wound to contract by gripping the wound edges and pulling them together.

The maturation phase refers to collagen fibers remodelation from type III to type I. This phase also called the remodeling phase starts around week 3 and can last up to 12 months. Collagenases are enzymes that break peptide bonds in collagen and eliminate the excess collagen, and wound contraction also begins. The resulting scar will never have 100% of the original strength of the wound and only about 80% of the tensile strength [15]. Once the proliferative phase has ended, many repairing cells will go under apoptosis. During the maturation phase, collagen fibers will slowly acquire a more parallel and closer disposition since water will be reabsorbed cross-linking of collagen reduces scar thickness and also makes the skin area of the wound strong.

3. Chronic wounds

Wounds generally heal in 4 to 6 weeks. Chronic wounds are those in which the healing process shows no response or a lower time rate of response over 4 or 6 weeks. Hypoxia, bacterial colonization, ischemia, reperfusion injury, altered cellular response, and collagen synthesis defects are factors that lead to impaired wound healing. Also, systemic conditions like diabetes, cancer, malnutrition, and smoking alter proper wound healing.

3.1 Biofilm

All open wounds, because they lack the protective covering of skin, are colonized within the first 24 hours, and they contain microorganisms that can be from the patient own's flora or from an exogen origin, for example, in hospitalized patients. Normally, these microbes are destroyed by the host's immune system. But, if microbes attach to the wound surface and proliferate, a biofilm will begin to develop, establish, and exhibit resistance to destruction by the host immune system and antimicrobials. At the wound site, microbes can proliferate exist under two distinct phenotypic states—planktonic (free-living) or biofilm (sessile/attached/aggregated). In the planktonic state, microbes can attach to a suitable surface (biotic or abiotic) and develop into polymicrobial biofilm aggregates. A biofilm can be described as a microbial colony encased in a polysaccharide matrix that can become attached to a wound surface. This biofilm, if proliferates to a critical point for instance more than 10^6 colony-forming units (CFUs) per tissue grams obtained by biopsy sample, can affect the healing potential due to the production of destructive enzymes and toxins, which can promote a chronic inflammatory state and also local or systemic infection. In the biofilm form, microbes have improved tolerance for antibiotics and host immune defenses [16, 17]. It is estimated that ~80% of the bacteria-producing chronic infections can form biofilms. During the process of biofilm formation, microorganisms can communicate with each other through quorum sensing. Quorum sensing regulates the metabolic activity of planktonic cells, and it can induce microbial biofilm formation and increased virulence [18]. Biofilm infection as defined *in vivo* based on criteria laid out by Parsek and Singh include i) aggregate embedded in extra polymeric substance (EPS) matrix; ii) adherence to a surface or each other; iii) persistent and localized infection; and iv) resistance to antimicrobial treatments [19].

3.1.1 Prevention and treatment of biofilm

Strategies to manage biofilm infection in wound care settings may be divided into three broad categories based on the aspect of the biofilm life cycle: adhesion inhibitors, biofilm maturation (communication) inhibitors, and promoters of disruption.

The management of invasive wound infection usually includes systemic antimicrobial therapy in combination with debridement. While debridement can be very powerful in debulking hostile biofilm aggregates, the lack of visualization of biofilm aggregates during debridement can inadvertently translocate bacteria into deeper tissue worsening local response and the patient's clinical condition [20, 21]. Surgical debridement converts biofilm bacteria to planktonic bacteria susceptible to antimicrobial therapy.

Other physical methods include nanomaterials (nanometer or submicron scale) which include nanoparticles made of metal or metal oxide. The membrane cell can be disrupted by the direct action of these nanoparticles or more indirectly by the liberation of free radicals or antiseptical drugs carried by liposomes.

Magnetic nanoparticles can be used to produce irreversible biofilm disruption (such as $\gamma Fe2O3$ maghemite or $Fe3O4$ magnetite nanoparticles) [22].

Nonthermal or atmospheric cold plasma (ACP) another physical alternative involves the generation of photons, electrons, neutrons, and protons when exposed to the constant supply of energy to a gas. The anti-biofilm effects of ACP are thought to be due to the generation of reactive oxygen species (ROS) and nitrogen species (RNS) (including organic radicals) [23].

Chemical methods are included in traditional strategies against biofilm. Silver-based management: The ionic form of silver ($Ag +$) has shown effectivity against bacteria (including methicillin-resistant *S. aureus* (MRSA) and vancomycin-resistant Enterococci (VRE)), viruses, and fungi in planktonic form. Silver-based wound dressings help to control wound biofilm by a controlled liberation of silver ions. This mode of antiseptic delivering to the wound bed is less cytotoxic for the growing cells which can proliferate in a less toxic environment. Silver nanoparticles are less toxic than ionic silver [24, 25].

Iodine is an antiseptic that impacts bacterial cells by multiple mechanisms. The neutral and lipophilic nature of iodine could enhance the penetration of this molecule into biofilms. It can kill planktonic cells rapidly and also inhibit mature biofilms of *Staphylococcus aureus* and *P. aeruginosa* when treated over an extended period. The cadexomer-iodine (CI) combinations that sequester iodine without limiting its inhibitory functions have been shown to have anti-biofilm effects [26].

Hypochlorous acid (HOCl) is known to rapidly eradicate pathogenic bacteria and is less toxic to mammalian cells than hydrogen peroxide. Some studies claimed that HOCl was bactericidal against Streptococcus strains but unable to disrupt biofilm, and *in vitro* efficacy against biofilm-forming Pseudomonas and Staphylococcus strains was demonstrated [27, 28].

Quorum sensing inhibitors: Bacteria can communicate with each other, exchange resistance mechanisms, and adapt their behavior collectively to their environment by a molecular phenomenon called quorum sensing (QS) that involves the synthesis and response to small molecules called autoinducers (AIs). QS drives the synthesis of virulence factors such as pyocyanin (P. aeruginosa), biofilm formation, and other activities. The inhibition of QS is called quorum quenching (QQ). Inhibitors with the QQ effect are numerous and range from natural (e.g., certain types of honey) [29, 30] to synthetic (e.g., furanone) [31].

Antimicrobial peptides: Antimicrobial compounds produced by honey microbiota Lactobacillaceae and Bacillaceae present in honey microbiota are known producers of antimicrobial compounds such as bacteriocins, surfactants, and siderophores. Environments rich in carbohydrates such honey and other beehive products facilitate the proliferation of lactic acid bacteria (LAB) and fructophilic lactic acid bacteria (FLAB) like Lactobacillus and Bacillus [32, 33].

Several LAB and FLAB isolated from pollen, honey, bee bread, and crop displayed antimicrobial activities against bee pathogens, foodborne, and multidrug-resistant human pathogens like multiple antibiotic-resistant Staphylococcus aureus, Staphylococcus epidermidis, Bacillus subtilis, and P. aeruginosa [34, 35].

LAB are well-recognized as the producers of the most active antifungal compounds against filamentous fungi, Aspergillus and Penicillium spp. and yeasts Saccharomyces, Candida, Kluyveromyces, Zygosaccharomyces, and Pichia spp. [36, 37].

Bacillales are another dominant order of Firmicutes-colonizing nectar and honey. Bacillales compound more than 50% of honey microbiota, ranging from 60 to 90% of all bacteria in the honey can. Bacillus strains produce gene-coded and nonribosomally synthesized antimicrobial peptides. Together with LAB, Bacillus spp. comprises an efficient factory that supplies honey with a broad range of antimicrobial compounds [38, 39].

A broad-spectrum activity has been demonstrated in most bacterial strains. There is also a reported bacteriostatic activity against several strains of Staphylococcus aureus, Staphylococcus epidermidis, Escherichia coli, Listeria monocytogenes, Pseudomonas aeruginosa, and Candida albicans [40, 41].

Honey and its derivates products are usually contaminated with fungus and yeast, which had become adapted to osmolarity changes due to sugar concentration variations and pH acidity changes. These contaminants also require to be resistant to other microbes antagonist interaction like bacteria.

To survive, fungi and yeast produce antimicrobial compounds, siderophores and surfactants, inhibiting bacterial growth and quorum sensing. This compounds result beneficial to human health [42–44].

The main antimicrobial mechanisms of action of dominant honey microbiota bacteria and fungi are: (a) the change of orientation of the peptide-lipid II complexes from parallel to perpendicular to the membrane, (b) the insertion of the C-terminal of the peptide into the cytoplasmic membrane, and (c) the formation of transmembrane, water-filled pore [45]. The second mode of action includes inhibition of cell wall biosynthesis through the binding to lipid II. Lipid II is the main transporter of a peptidoglycan subunit from the cell interior to the place of cell wall synthesis [46].

3.2 Pressure sores

Pressure ulcers are considered chronic cutaneous wounds caused by sustained pressure over time, deformation, friction, and rubbing. Usually, bed sores appear in bedridden patients who cannot move or change their body positioning by their own will or because their consciousness is altered. Besides, the healing process can be prolonged and slow because of malnutrition, anemia, and the use of vasopressor drugs or anti-inflammatory drugs like corticosteroids. These lesions are usually located in the skin and underlying tissues on bony prominences [47].

Pressure ulcers represent a major problem in the healthcare system, with great epidemiologic, economic, and socio-family impact [48]. Approximately 65% of pressure ulcers originate in hospital stays, affecting mostly patients above 65 years of age and producing a negative impact on their nutritional and metabolic status and alteration in bowel movement, mobility, cognitive, and perceptual capacity, and the skin barrier [49].

Hospital-acquired pressure ulcers are considered one of the healthcare quality indicators, and it is estimated that 55–77% of cases could be avoided [50]. Early detection of patients at risk of developing a pressure injury needs to be implemented from the time of admission. Age, incontinence, and body mass index (BMI) are well-established risk factors. Having a BMI < 19 kg/square meter and an age above 65 are the most significant risk factors [51, 52]. Incidence is lower in overweight (BMI 25–30 kg/square meter) and obese (BMI >30 kg/square meter) patients [53].

The Braden Scale assesses the risk of developing a pressure injury based on criteria related to activity, mobility, skin moisture, nutritional status, friction, and rubbing, as well as the ability to feel pain and discomfort in different parts of the body as a result of pressure [54, 55].

Charlson's Comorbidity Index permits the assessment of comorbidity factors related to the mortality risk of hospitalized patients. Patients with vascular disease, diabetes, chronic renal or pulmonary disease, and cancer are at a higher risk of developing pressure ulcers [56, 57].

Another important aspect of risk evaluation can be the categorization by related risk groups, for example, clustering groups of surgical patients and nonsurgical patients during hospitalization [58]. Surgical patients are at higher risk of developing pressure ulcers due to factors such as longer immobilization, anesthesia and/or surgery, and pre-existing medical conditions [59, 60].

3.2.1 Pressure sore prevention

The patient's skin evaluation is fundamental to pressure injury prevention, classification/diagnosis, and treatment. All inpatients should have a skin assessment to determine their general condition and identify factors that increase the risk for pressure ulcer development. This examination should be performed from the initial patient's admission throughout his route at different hospital units: Urgency, Intensive care, General Unit up to discharge. The initial condition of the patient's skin at the admission first initial exam, like thin and fragile skin or senile purpura, is the most important early indicator of the skin's reaction to pressure exposure and the continuing risk of pressure injury [61].

The complete skin assessment should include a general visual check of the entire body's skin and mucosae surface and identify any characteristics indicative of pressure damage. This first signs include redness, local pain, and blistering. Caregivers should frequently monitor and check the skin beneath dressings, prostheses, and devices, and also areas of localized heat, skin breakdown, edema, areas of redness that do not blanch, and induration of the wound.

Bony prominence areas, which are at an increased risk for pressure injury due to pressure, friction, and shearing forces, should be protected from the very beginning of the admission. High-risk areas include the sacrum, heels, elbows, wrists, temporal region of skull, ears, shoulders, back of the head (especially in children less than 36 months of age), knees, and toes.

The patient's baseline level of mobility and current level of mobility should be considered to plan an accurate strategy to prevent bed sores. For patients who can move independently or assist in moving themselves, it is recommended that they be encouraged and/or assisted to change their position regularly, either in bed or out of bed (if able). The selection of an appropriate support surface should take into consideration factors such as the individual's level of mobility within the bed, the patient's weight, and the need for microclimate control. When a patient's voluntary movement is altered, changing body position every 2 hours is recommended.

Increased moisture on the skin or excessive dryness can exacerbate pressure injury development due to the risk of skin breakdown and altered skin integrity.

To maintain the skin clean and dry, the skin must be daily cleaned to remove unwanted substances without rubbing or frictioning the patient's skin; pH neutral or slightly acidic skin cleanser (pH 4–7) is recommended and alkaline products (pH >7) should be avoided. To avoid skin dryness, a fragrance-free moisturizer is recommended.

Caregivers should protect the patient's skin from incontinence. It is recommended to clean skin promptly after episodes of incontinence and choose appropriately sized incontinence products for maximum absorption. To create a physical barrier between the skin and fecal or urine irritants the use of barrier creams is recommended to prevent the skin breakdown.

Patients in pain are at an increased risk of pressure injury. Analgesia should be adequate during patient hospitalization, and analgesic rescue must be performed if a patient undergoes major mobilization that could expose injured skin areas and cause pain.

Malnourished patients are at increased risk of pressure injury development. Hydration and nutritional support should keep the nitrogen balance and serum albumin levels according to renal and liver function to maintain adequate skin integrity and hydration.

3.2.2 Pressure sores classification and wound management

Pressure injury stages and according to treatment indication.

Pressure injury staging or classification describes the extent of skin and tissue damage. Staging a pressure injury is essential for the development and implementation of a management plan.

Stage 1: Localized área of nonblanchable erythema of intact skin: protects the skin to prevent further injury. Dressing selection: silicone adhesive, nonadherent foam, or transparent hydrocolloid adhesive dressing.

Stage 2 Partial thickness skin loss relieves pressure and protects the wound from further trauma and contamination. Dressing: silicone adhesive or nonadherent foam.

Stage 3 Full-thickness skin loss: relieve pressure and protect the wound from further trauma and contamination. Dressing: alginate, hydrogel, hydrofiber, or silicone dressing.

In the treatment of cavitated wounds, different therapeutic options must focus on stimulating granulation at the dermis level to achieve coverage of underlying structures such as muscles, bones, or tendons.

A critical point from the therapeutic perspective in order to prevent and or treat pressure sores is the capability of healthcare givers to recognize early stages of pressure sores among the different types of skin colors and be skilled in the adequate selection of intelligent dressing and devices. It is also important to prevent fecal or urine prolonged contact to patient's skin to avoid further irritation and or contamination.

Currently, calcium alginate is frequently used, as well as negative pressure systems or grafts for cavitated wounds.

The choice will depend on the availability of financial and human resources and hospital infrastructure. Advanced wound care, proper diagnostic understanding in each case, and management of the patient and his/her family context are all necessary conditions to achieve good clinical results. The insufficient availability of skilled human resources as well as the high cost of materials required for advanced care of wounds in developing countries call for the development of easy-to-implement, reproducible, and economical strategies that can be supervised by telemedicine.

During the last decade, the therapeutic approach to cavitated pressure ulcers has made great progress with the advancement of new technologies such as negative pressure, which promotes the filling of deep wounds [61]. Negative pressure cannot be applied in the presence of cancer tissue in the wound, to untreated osteomyelitis, in wounds with organ or vessel exposure, or to areas with poor blood flow. In some trials, its use in infected grade IV sacral ulcers has proved advantageous in terms of recovery time as compared to conventional treatments, achieving a recovery time duration of approximately 6 weeks [62]. However, in developing countries, this type of device continues to have a high cost, not only because of the elements required but also because of the skilled staff required for correct application and removal.

Other options are flaps and grafts close to the anal region, but these may lead to complications such as infection and necrosis. Usually, the sacral and perianal regions create difficulties in the adjustment and fixation of any device. Combined techniques and strategies, such as flaps plus a negative pressure system, tend to work well, although they also imply greater therapeutic demands [63].

A low-cost option for the treatment of cavitated wounds is calcium alginate, alone or in combination with micronized silver, with high absorptive and hemostatic capacity and some bactericidal properties. Calcium alginate may be applied in rope dressings that are easily introduced in tunneled wounds and can easily integrate naturally into the tissue. The dressings can be changed every 72 hours. When this product dissolves, it releases a gel-type discharge from the wound bed with a special smell that may be mistaken for infection. It requires a secondary dressing [64].

3.3 Skin genetic disorders: epidermolysis bullosa

Epidermolysis bullosa (EB) is a hereditary genetic skin disorder, classified as a type of genodermatosis, which causes severe, chronic skin blisters associated with painful and potentially life-threatening complications. Currently, there is no effective therapy or cure for EB.

Epidermolysis bullosa (EB) is a rare disease that affects 1 in 40,000 to 50,000 newborns worldwide.

Epidermolysis bullosa is a genetic disease caused by a modification, decrease, or absence of structural proteins in the skin. It is caused by a mutation of the gene that codes for that protein. Genetics has identified over 20 genes that are responsible for this disease, which makes it difficult to cure. It is characterized by lesions and blisters on the skin and mucous membranes (oral cavity, nasal cavity, pharynx, esophagus, digestive tract, respiratory tract, genitourinary tract, perianal area, and conjunctivae) that occur after minimal trauma or even spontaneously.

In epidermolysis bullosa (EB), blisters with serum-hematic fluid content appear on the skin. When these blisters burst, burn-like lesions occur, which then progress into scars and skin retractions that lead to functional disabilities such as pseudosyndactyly, among others. Lesions also occur on mucous membranes causing complications at the gastrointestinal tract level, resulting in malnutrition which further complicates healing.

EB is characterized by a pronounced skin and mucosal frailty that triggers blister and ulcer formation in response to minor trauma. The main affected sites are those exposed to frequent friction and pressure.

Perioral tissues, lip and cheek skin, and mucous membranes are especially affected by constant exposure to the "trauma" of chewing. Ulcers in the perioral region due to minimal trauma during chewing result in scars and tissue thickness which, when retracted, cause microstomia, thus making it difficult to open the mouth.

Localized blisters are the most common feature in all types of EB. However, depending on the EB subtype, the extent of involvement will vary from small vesicles that heal without causing damage to extensive lesions that lead to scarring and subsequent oral deformity, among others.

Vesicles have been identified in 92% of patients with recessive dystrophic EB, the tongue being the most affected area causing lingual depapillation.

Vesicle scarring leads to ankyloglossia (tongue sticking to the floor of the mouth) and obliteration of the buccal vestibule, which results in difficulty in chewing and suction movements. In addition, the lack of normal mobility of the oral cavity causes bone and dental development problems in the maxillae.

As a result of scar tissue formation in the mucosal lining and the skin of the lip region, especially in the commissures, there is a reduced inter-incisive buccal opening.

The reason for the delay in maxillary and consequently craniofacial growth is associated with impaired oral and masticatory movements and malnutrition due to reduced and inadequate food intake.

The formation of intraoral scar adhesions results in the collapse of the maxillae, palatal atrophy, retained teeth, dental misalignment, and dental crowding. Therefore, patients suffer from malocclusion.

Microstomia (limited mouth opening) and ankyloglossia cause muscle function alterations and subsequent craniofacial and dental growth impairment. Although the full spectrum of clinical manifestations is heterogeneous, blistering, pruritus, skin erosions, atrophic scarring, hyperkeratosis, and ulcers are the main cutaneous expression of the disease.

There are three main types of EB plus Kindler's syndrome:

EB simplex (EBS) comprises all subtypes of EB that have mechanical fragility and blistering localized in the epidermis. There are two subgroups: suprabasal and basal depending on the histopathological site where the blisters originate in the epidermis. The most frequent mutations are found in the genes coding for the proteins Cytokeratin 5 and 14, Plectin, and Transglutaminase 5, among others. Inheritance is mainly autosomal dominant, but there are rare subtypes of recessive inheritance. There are generalized forms with lesions appearing at birth scattered all over the body and others localized mainly on the hands and feet. All EBS forms are exacerbated in hot and humid environments where increased sweat production contributes to blister formation. There is no growth delay, anemia, or esophageal lesions in this type of EB.

Junctional EB (JEB) includes all EB subtypes in which blisters are found in the lamina lucida located in the area where the basement membrane joins the epidermis and dermis and overlying the lamina densa. There are two subgroups: generalized and localized. The most frequent mutations are observed in the genes coding for proteins Laminin 332, Collagen type VIII, and Integrin $\alpha 6\beta 4$. Inheritance is autosomal recessive. JEB is a severe form of the disease. It is characterized by widely distributed blisters, hyperplastic granulation tissue in the perioral, perinasal, and nail regions, or blistering sites. It also involves the mucous membrane of the mouth (intraoral vesicles), larynx, bronchi, esophagus, rectum, and vagina. Extensive denuded areas are observed at friction sites. The combination of chronic infections and iron depletion can lead to chronic anemia; it can be associated with growth delay, malnutrition, and involvement of the oral cavity and dentition.

Dystrophic EB (DEB) includes all DEB subtypes where blistering occurs within the upper dermis just below the lamina densa of the skin. There are two subgroups: dominant (DDEB) and recessive (RDEB). All mutations of this EB type are seen in the gene coding for Collagen type VII protein. Inheritance can be dominant or recessive, with recessive inheritance being more severe. Some DDEB forms may involve hands, feet, knees, nails, and other organs, whereas RDEB shows extensive blisters and lesions all over the body, nail loss, anemia, growth delay, pseudosyndactyly (fusion of the fingers or hands in the form of a cocoon), and wounds in the cornea, mouth, esophagus, and bowel. There is a high probability of bacterial colonization and recurrent infections in wounds. Also, there is a risk of developing skin cancer (spinocellular carcinoma).

In the Kindler syndrome (KS), of autosomal recessive inheritance, mutations are observed in the gene coding for the Kindlin 1 protein. It is the least common form of EB, and it is difficult to diagnose. It is characterized by the presence of phenotypic features unique to EB: photosensitivity, poikiloderma, and scarring.

New scientific advances despite their high cost and complex clinical applicability open up new pads for future treatments. This includes gene therapy, protein replacement therapy, cell therapy with allogeneic fibroblasts, mesenchymal stromal cells, and gene editing/engineering. Tissue engineering materials that mimic skin structure are also now available. However everyday patients′ reality is much closer to standard treatments like nonadherent dressing and ointments.

In addition to this, many intrinsic factors adversely affect patient's healing process that includes anemia, malnutrition, infection, and itching.

The constant skin detachment leaves numerous body denudated areas that undergo the defective repair process. Incidental pain, pruritus and scarring, and local infections increase the incidence of squamous cell carcinoma, which represents the main cause of death in young adults with RDEB.

Goertz et al. described a solidifying gel that dissolves at room temperature and hardens to a gel consistency at normal body temperature or above. This gel reduces incidental and spontaneous pain.

Another new smart dressing is a gelling dendrite dressing based on hydrogel, which solidifies in few minutes, ensure hemostasis, and provide a balanced moisture environment. It can be dissolved any time [65].

Antibacterial gel dressings based on chitosan (Opticell Ag+) have recently been introduced, which provide a moist, adaptable, highly absorbable antimicrobial dressing to reduce dressing changes and alleviate pain [66].

Honey impregnated dressings and ointments are effective in both the treatment of chronic wounds and in reducing the biological load [67].

Cutimed Sorbact dressings remove bacteria through hydrophobic interactions. They are coated with a fatty acid derivative that attracts bacteria to the dressing, where they are bound [68].

Dressings containing polyhexanide, such as Suprasorb X1 PHMB (Activa Healthcare, Lohmann & Rauscher, UK), provide antimicrobial treatment for critically colonized and infected wounds, and they are recommended for long-term application [69].

The polymer membrane dressing (PolyMem, Ferris, OH, USA) contains a cleaning agent (surfactant), which reduces the biological load and allows the healing of resistant wounds. Polymeric membrane dressings have the advantage of being "self-contained" without the need for a nonadherent primary or secondary dressing to protect or manage exudation [70].

Infected or critically colonized wounds require more frequent dressing changes. Another important aspect to consider is that an ideal dressing should be also usable on oral, ophthalmic, or genital mucosa, especially on this kind of patient.

3.4 Domestic burns

Burns are a global public health problem, accounting for an estimated 180,000 deaths annually. In 2004, nearly 11 million people worldwide were burned severely enough to require medical attention. In India, over 1,000,000 people are moderately or severely burnt every year. The majority of these occur in low- and middle-income countries, and almost two-thirds occur in the WHO African and South-East Asia regions. Burns are among the leading causes of disability-adjusted life-years (DALYs) lost in low- and middle-income countries. Nonfatal burn injuries are a leading cause of morbidity, occur mainly in the home and workplace, and are preventable. Nonfatal burns are a leading cause of morbidity, including prolonged hospitalization, disfigurement, and disability, often with resulting stigma and rejection [71].

A burn is an injury to the skin or other organic tissue primarily caused by heat or due to radiation, radioactivity, electricity, friction, or contact with chemicals. Thermal (heat) burns occur when some or all of the cells in the skin or other tissues are destroyed by: hot liquids (scalds), hot solids (contact burns), or flame (flame burns).

Females have slightly higher rates of death from burns. Open-fire cooking can ignite loose clothing of women suffering major burns areas. Unsafe cookstoves and neglected parent-control environments put also children at risk when they are nearby [71].

Open flames used for heating and lighting also pose risks, and self-directed or interpersonal violence is also a factor. In Argentina, since 2016 there has been an increased number of domestic violence cases against women perpetrated by male partners. These attacks are characterized by flammable body burn injuries. Usually, men spray alcohol against the victim's body and then light it up with any kind of flammable object, cigars, matches, for instance. In 2016 were registered 256 total cases of femicides, 7% of them were burn injuries [72].

It is mandatory to find accessible and affordable treatments for domestic burns in developing countries. This should be accompanied by educational first-aid strategies. Medically tested natural products like beeswax and cod liver oil combined ointment, constitute a great treatment option for domestic burns.

4. Curefini™ natural ointment

Curefini™ is an OTC product approved by the FDA and is currently in use in pediatric patients with recessive dystrophic epidermolysis bullosa (RDEB). Curefini's main components are petrolatum, sunflower oil, cod liver oil, sweet almond oil, and medical beeswax. Curefini™ polyunsaturated fatty acids (PUFAs) component gives anti-inflammatory properties that facilitate a controlled wound inflammatory environment, enhance reepithelialization and granulation processes, and allow for autolytic debridement and gentile mechanical detritus removal on each gauze change.

Curefini™ ointment is a formula enriched with PUFAs, omega 3, and omega 6, contained in its pure source oils. PUFAs reduce the local inflammatory process in the wound by inhibiting the production of inflammatory eicosanoids and competitive inhibition of the formation of arachidonic acid. PUFAs also have antibacterial action [8].

Sunflower seed oil contains large amounts of linoleic acid, which has anti-inflammatory and antibacterial properties and promotes the restoration of the skin barrier [73].

Sweet almond oil is restorative to the skin barrier and has anti-pruritic properties [74].

Beeswax is known for its healing, antibacterial, and anti-inflammatory properties [4, 5].

The combination of cod liver oil and medicinal beeswax has been successfully used in wound treatment in veterinary medicine [9, 10].

Curefini® has been previously tested on a porcine model for second-degree burns and showed the ability to control the inflammatory process and promote the development of a more-resistant skin layer in the affected area [75].

4.1 Polyunsaturated fatty acids, PUFAs

Adequate skin response to an injury depends not only on prior skin health but also on many crucial factors of them mentioned before, like nutritional status and optimum tissue oxygen supply, and an equally important factor is a balanced inflammation microenvironment. In the same way, as heat is needed to cook a meal, in the tissue regeneration process, the immune system and free oxygen radicals should be balanced and controlled in time and intensity to restore damaged structures. Bioactive lipids and cytokines regulate the skin's immune system. They can initiate an immune response with controlled inflammation, followed by efficient resolution.

Unsaturated fatty acids such as linoleic acid (LA), α-linolenic acid, and oleic acid, and most of their bioactive products have shown an effective role as a topical treatment of chronic skin wounds. Their effect, when the treatment starts at day 0, has been observed mainly in the inflammatory phase of the wound healing process. Thanks to their bioactive properties, unsaturated fatty acids (monounsaturated fatty acids (MUFAs) and polyunsaturated fatty acids (PUFAs)) enhance the wound healing process in chronic wounds.

A fatty acid chain is monounsaturated if it contains one double covalent bond between the carbon atoms and polyunsaturated if it contains more than one double covalent bond. According to the length of the chain, they can be classified into short chain (2–6 carbon atoms), médium chain (13–21), and long chain (more than 22). The metabolic rate of fatty acid depends on the number of carbons and double bonds, the location of the double-bond position, and its geometric configuration (cis or trans) [76]. The double-bond position in the chain, defined by the number of carbon atoms from the terminal methyl group (omega or ω) (H3C), determines the "metabolic family" to which the fatty acid belongs.

The human body cannot synthesize some PUFAs which are those from the omega-3 (ω-3 or n-3) and omega-6 (ω-6 or n-6) families. They have the first double bond in the third and the sixth carbon, respectively, counting from the terminal methyl group. α-Linolenic acid (ALA) belongs to the ω-3 family and is an important precursor of other fatty acids with bioactive properties, such as eicosapentaenoic acid (EPA) and docosahexaenoic acid (DHA). On the contrary, linoleic acid (LA) belongs to the ω-6 family and is a precursor of arachidonic acid (AA), an important fatty acid with several metabolic effects, since it originates other eicosanoids from the ω-6 family, such as leukotriene, prostaglandins, and thromboxane.

Even though the human body can produce saturated and unsaturated fatty acids from carbohydrates and proteins, it is not able to synthesize ALA and LA [77].

PUFAs can regulate and modulate inflammation and immune response due to eicosanoid production [78]. Eicosanoids are originated from omega-6 and -3 PUFA metabolism, such as prostaglandins, leukotrienes, thromboxanes, and lipoxins, which modulate the inflammatory response unevenly [79].

Eicosanoids from omega-6 PUFA metabolism, produced by AA oxidation, are potent inflammatory mediators, involved in infection, inflammation, tissue damage, immune system modulation, and platelet aggregation. On the other hand, with a predominant anti-inflammatory role, ALA (omega-3 family) can be converted into EPA and DHA, which compete with AA for the enzymatic pathways of cyclooxygenase (COX) and lipoxygenase (LOX), also giving rise to eicosanoids [80].

During skin wound healing, keratinocyte and fibroblast produce PGE2, which has a pro-inflammatory action, promoting vasodilatation, cell proliferation, and modulating immune response [81, 82]. PGE2 also regulates fibroblast migration and collagen contraction and has an anti-fibrotic effect by increasing the expression of matrix metalloproteinase (MMP) 2 and 9, which are type IV collagenase found at elevated levels in chronic wounds and inhibiting TGF-β1-induced collagen synthesis of dermal fibroblasts [83]. PGE2 is also involved in keratinocyte proliferation and differentiation, the chemotaxis of keratinocytes, and the modulation of dermal fibroblasts [84].

The metabolism of AA (omega 6 family) *via* COX and LOX pathways generates proinflammatory products, while the metabolism of ALA and LA (omega-3 family) results in anti-inflammatory action products.

Skin macrophages have nuclear receptors called peroxisome proliferator-activated receptors (PPARs) that can regulate their anti-inflammatory state reducing the production of pro-inflammatory cytokines [85].

PUFAs such as 15-hydroxy-eicosatetraenoic acids, 13-hydroxy-octadecadienoic acid, 15-doxy-Δ-12, and 14-prostaglandin J2 (15d-PGJ2) can activate PPAR-γ receptor reducing the levels of proinflammatory cytokines such as IL-1β, TNF-α, and IL-6 in the wound, and slightly increase the levels of pro healing growth factors, such as VEGF, IGF, and TGF-β1 [86].

Topical treatment with polyunsaturated fatty acids (PUFAs) and fatty acids' analogs as well as their receptors' agonists has positive effects in wound healing. They have become an interesting treatment option to enhance the wound healing process.

Cod liver oil is rich in the following PUFAs: eicosapentaenoic acid (EPA) and docosahexaenoic acid (DHA), sunflower, and almond oil contain linoleic acid (LA) and gamma-linolenic acid (GLA) all of them present in Curefini™ formula.

4.2 Beeswax

Beeswax is as a complex lipid with organic constituents displayed in a liquid form, produced by the specialized wax glands of the bees. Beeswax changes its glass-clear and colorless appearance after worker bees masticates it and then becomes more brownish with the incorporation of pollen oils and propolis.

Its chemical composition is represented by a combination of almost 300 constituents such as fatty acid esters (approximately 67%), free fatty acids (12–14%), diesters, hydrocarbons (among 12–16%), fatty alcohol (approximately 1%), and exogenous substances (propolis and pollen residues, and a reduced number of floral constituents) [87].

Beeswax also has flavonoids and antioxidants which have positive effects on the wound healing process when applied topically regulating the production of cytokines by skin inflammatory cells [88, 89].

Virgin beeswax was observed to have antibacterial inhibition against different types of bacteria and also *C. albicans* yeast. The efficiency of beeswax can be detected against some Gram-positive bacteria, including *S. epidermidis*, *Streptococcus pyogenes* (*S. pyogenes*), and S. aureus but also on *P. aeruginosa*, *B. subtilis*, and *E. coli* as part of Gram-negative bacteria. A smaller impact was shown among *C. albicans* while against *Proteus mirabilis* and *Salmonella typhimurium*, beeswax proved to have no effect. The synergy between honey, olive oil, and beeswax was demonstrated to have an important impact on *C. albicans* and *S. aureus* [5, 90].

5. Case reports: clinical experience with Curefini™ ointment

5.1 Case 1: grade IV sacral sore treated with Curefini™ ointment

The present Case Report has been published at Case Reports in Dermatological Medicine Journal Open Access publisher CC-By 4.0 license [91].

Fifty-seven-year-old male patient with a history of hypertension, dyslipidemia, obesity, and coronary disease underwent double bypass surgery, with a 27-day-post op interment period during which he developed a sacral pressure ulcer with asymmetrical compromise on both sides of the intergluteal cleft, grade III, and grade IV pressure ulcers on the left and right sides, respectively, according to the NUAP/EUAP classification [92]. The patient's photographs showed a sacral ulcer involving the intergluteal cleft in a butterfly shape, with necrotic tissue 10 cm diameter on each side. Involvement of the deep tissues was not evident at first sight, but the left side seemed to have more superficial involvement (**Figure 1a**). The treating physician decided to initiate autolytic debridement with chloramphenicol/collagenase ointment plus secondary gauze dressing with petrolatum, cleaning with polyhexanide solution twice daily. After 10 days of evolution, the first surgical toilette was performed with debridement of necrotized tissue; no swabs or culture specimens were collected. An ulcerative granulation bed was observed in the deep dermis with active borders and some brownish-red areas on both sides. No local erythema, temperature, or purulent discharge was observed (**Figure 1b**). The patient received prophylactic antibiotic therapy with ciprofloxacin 500 mg PO bid for five days and continued with chloramphenicol/collagenase treatment at home. After 18 days of chloramphenicol/collagenase local treatment, an ulcerative bed was involving sectors up to the hypodermis with fatty tissue exposure and cavitation at the right median and paramedian sacral level, reaching the muscular plane with no evident bone exposure (**Figure 1c**). Neither the probe to bone nor the lumbosacral IRM was performed. A fetid, purulent discharge was found on the gauze. No fever or shivering was reported. It was decided to schedule a new surgical toilette which was performed on day 20.

On day 20, a second surgical debridement was performed. Physical examination showed a left-side grade III pressure sore with granulation tissue at the wound bed and active reepithelialization borders; on the right side, a cavitated grade IV sore was observed, with hypodermis fat exposure, increased depth of the cavity with no bone exposure, and the negative probe to bone test (**Figure 1d**).

Following the debridement, according to the patient's medical report, no soft tissue culture specimens were collected and empiric systemic antibiotic treatment was indicated:

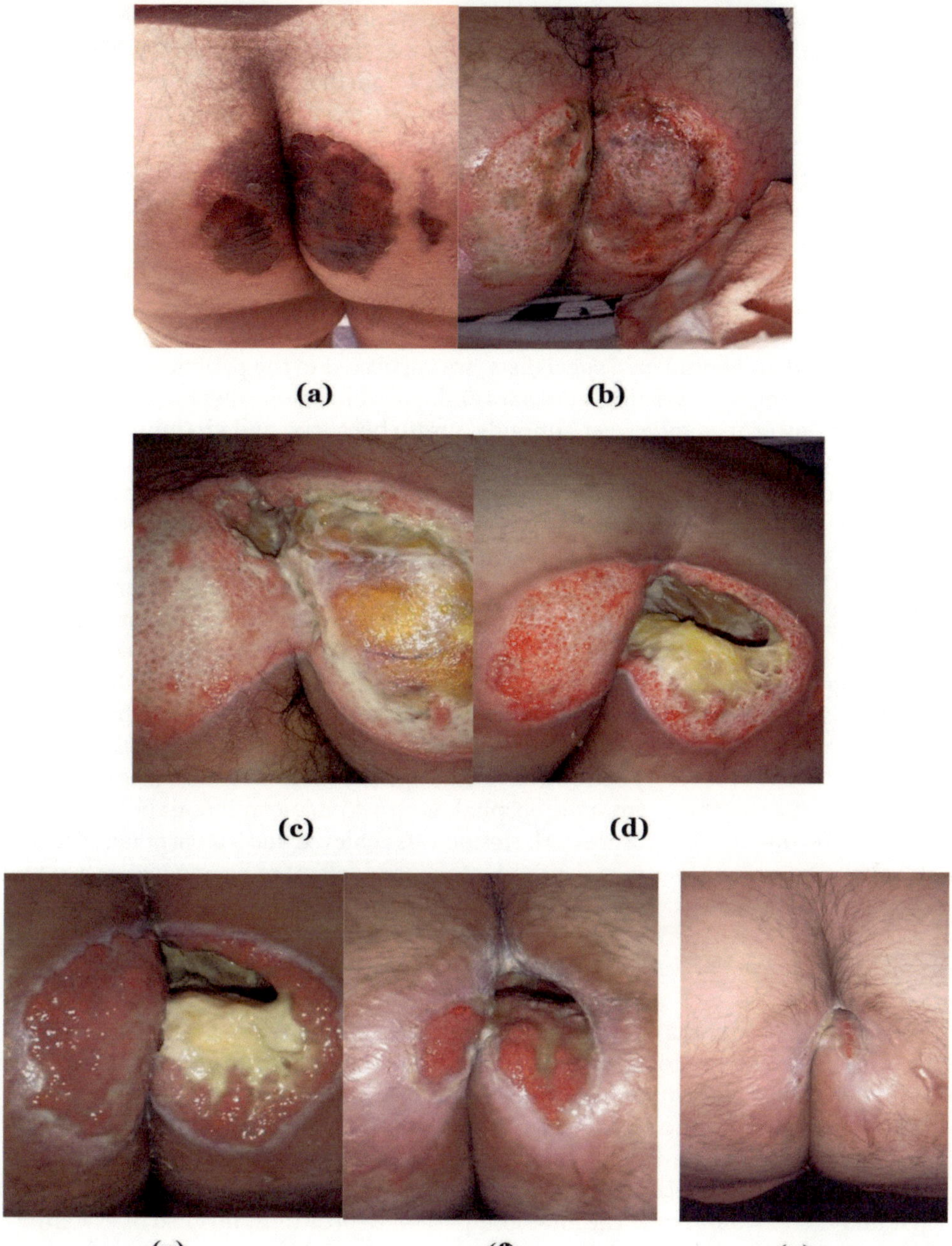

Figure 1.
Wound progression before Curefini®treatment. (a) Day 1: the patient is discharged from hospital after 27 days. Chloramphenicol/collagen is initiated as local treatment. (b) Day 10: the first surgical debridement is performed. Chloramphenicol/collagenase local treatment continues. (c) Day 18: topical treatment with chloramphenicol/ collagenase continues. (d) Day 20: the second surgical debridement is performed, initiated with iodoform gauze, antibiotics PO, and oral opioids. Wound progression with Curefini®treatment. (e) Day 42: after 12 days of Curefini®local treatment. (f) Day 60: after 30 days of Curefini®. (g) Day 75: after 45 days of Curefini® local treatment.

amoxicillin/clavulanate 875/125 mg PO bid for 10 days and local treatment with iodoform gauze dressings inside the cavity, with the indication to change every 48 hours. The patient complained of an increase in local pain and received tramadol 50 mg PO every 8 hours.

On day 30, the treating physician proposed a skin graft; however, the patient decided to get a second opinion from a wound care specialist. At this point, the authors contacted the patient for the first time. During the interview, the patient was said to be concerned about the evolution of the wound, was said to be unwilling to have a surgical procedure and later hospitalization since his home is located 500 km away from the hospital, and had reported continuous pain and impossibility to sit or find a good position to sleep and regarding his mood, and he seemed to be more depressed.

As independent wound care specialists, we suggested to the patient and his family a new local treatment consisting of gauzes embedded in an ointment made of cod liver oil, sunflower seed oil, sweet almonds, virgin beeswax, and vitamins A and D, Curefini™. One month after being discharged from the Coronary Unit, the patient started topical treatment with a sterile gauze lubricated with Curefini™, filling the cavity. The surface of both buttocks was covered with sterile gauze embedded with Curefini™, with the indication to change it every 12 hours.

On day 42, after 12 days of topical treatment with natural ointment, pain improved significantly, oral opioids were tapered until total discontinuation, and the patient tolerated the sitting position for brief periods. After Curefini™ local treatment was established, the granulation and reepithelialization processes were reactivated and hypodermis fat tissue was covered with new skin (**Figure 1e**). On day 60, after 30 days of treatment with Curefini, a 50% reduction in transverse wound diameter was observed, along with granulation tissue in the cavity bed and advanced epithelialization border (**Figure 1f**).

On day 75 after discharge from the hospital, with a 45-day-continuous period of Curefini™ treatment, 95% closure of the lesion was achieved and a large crater ulcer was replaced by a transverse fissure in the right intergluteal fold of only 4 cm in depth, which continued to be treated locally with a gauze impregnated with the ointment (**Figure 1g**).

At this point, the patient starts to walk without difficulty or pain, maintains a correct diet with normal bowel movements, has restorative sleep, and shows a tranquil mood.

Control after 3 months of Curefini™ treatment showed the cavitation was filled, although a certain degree of scar retraction is observed in the surrounding tissue with some hypertrophy. No remnant fistula trajectory or tunnel is observed.

This case highlights the therapeutic efficacy of treatment with a natural ingredient-based product on cavitated wounds that help to reduce pain and promote granulation and reepithelialization of the skin. We remark that despite the location of a cavitated sacral wound near the anus and patient depositions, no local or systemic infectious events were detected. Its use should be considered a treatment of choice in developing countries where the distance from the hospital, economic limitations, and lack of wound care specialists are key conditions for a successful outcome.

Curefini™ can be used for prevention as a barrier to avoid shearing, friction, and also on early stages pressure sores.

5.2 Case report 2: second-degree burn treated with Curefini™ ointment

The present case report was published in Wounds International Journal [11].

A 51-year-old female patient with no clinical records of health problems consulted for a boiling oil scald of her skilled hand that occurred 4 days previous to consultation. She felt an extreme burning and piercing pain. The color of the skin was initially red, developing a blister of serous content with a size of 7 per 4 centimeters in diameter red on the base and purple surrounding it (**Figure 2a**). At home, the patient placed her hand under cool running water to relieve pain and covered her hand with a gauze dressing. The blister broke spontaneously, draining part of its content (**Figure 2b**). She was immediately treated in the Emergency Room where the blister was debrided and the area was treated with silver sulfadiazine and lidocaine, sterile gauze, and gauze dressing. Spontaneous and incidental pain did not improve, so the patient decided to seek cross-consultation with the Dermatology Department 4 days after the incident. On physical examination, there were no infection findings, the wound bed was intense red and there was no tissue fragment or cellular detritus. The mobility of fingers was limited by pain (**Figure 2c**).

The treatment protocol is decided to be modified initiating treatment with Curefini TM ointment extending a thick layer on a sterile dressing to therefore apply it over the area affected covering the hand and forearm with PVC cling film.

A daily wound treatment course was prescribed. About 24 hours after the onset of treatment, the patient experienced significant pain relief which allowed her to sleep and started to move her hand. After subsequent medical control visits, the attenuation of intense red color, which for the days turned pink, could be observed (**Figure 2d–f**).

At 3 weeks, reepithelization of wound healing (**Figure 2g**) was achieved by 80%. Wound healing treatment continued for 4 weeks exactly according to the onset of treatment, since the patient still referred skin sensitization to the topical exposure of her hand exposed, daily rubs, and climate change. At the end of the period, skin showed normal thickness reporting no scarring, retractions, or hypersensitivity. The opening and closing hand functions of fingers as well as the closed fist function were achieved (**Figure 2h**). During the whole process, the wound did not get infected.

Curefini^{TM} ointment reported the following:

- Decreased spontaneous and incidental pain
- Favored reepithelization.
- Provided natural components with anti-inflammatory and antibacterial capacity offering an alternative to the common use of products containing antibiotics or corticosteroids.
- Resulted effective in wound healing treatment for second-grade burns in human adults. It was previously tested on a porcine model for second-degree burns showing an ability to control the inflammatory process and promote the development of a more resistant skin layer in the affected área
- Maintained skin moisturized and decreased desiccation. Curefini TM has the potential to be used in other inflammatory and desquamated skin states.
- Stimulated the occlusive healing, plus the impermeability of the secondary dressing, PVC cling film by avoiding direct contact with air, and low tension oxygen tension on skin surface [93].

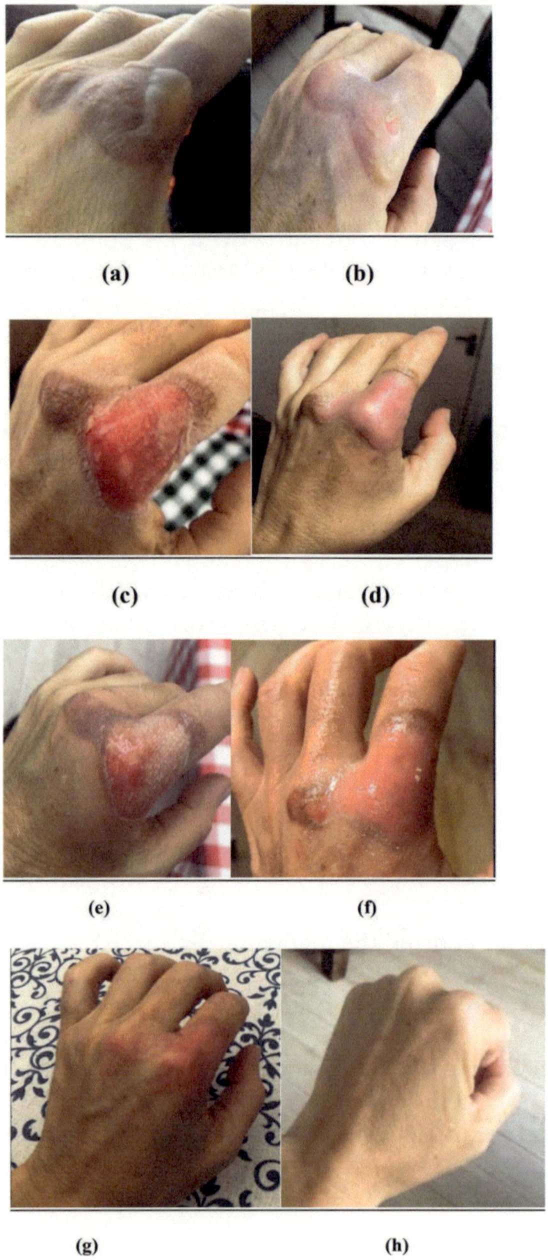

Figure 2.
(a) and (b) The first six hours after burn injury with boiling oil, a serum-filled blister measuring 7×4cm was formed, with purple tissue surrounding it. Initially, the patient had placed her hand under cool running water to relieve pain and covered her hand with a gauze dressing. The blister broke spontaneously. She was then treated in the emergency department, where the blister was debrided (c) and the wound was treated with silver sulfadiazine and lidocaine, with a sterile gauze dressing. Her spontaneous and incidental pain did not improve, so she decided to seek a consultation with the dermatology department on day 4 (d) On physical examination, there were no signs of infection, the wound bed was bright red and there were no tissue fragments or cellular debris. The mobility of her fingers was limited by pain. The treatment protocol was changed to Curefini™ ointment, using a thick layer on a sterile dressing on the wound, and a secondary dressing of PVC clingfilm. (e–g) At 3 weeks, 80% of the wound had reepithelised. At the end of 6 weeks (h), the skin showed normal thickness reporting no scarring, contracture or hypersensitivity. She had recovered her range of motion and could flex and extend her finger and close her hand into a fist. No infection developed at any stage.

5.3 Case 3: long-term follow-up study of nine (9) years of a female patient diagnosed at birth with recessive dystrophic epidermolysis bullosa, under continuous treatment of skin and oral mucosa with a topical ointment elaborated with petrolatum, cod liver oil, virgin bee wax, sunflower oil, sweet almond oil, and vitamins a, D, and E. Assessment of skin development and quality of life (QoL) impact

Background: Full-term AGA, born by cesarean section on 10/25/11, Apgar 8/10. Clinical diagnosis: Recessive Dystrophic Epidermolysis Bullosa (RDEB) with aplasia cutis in feet, confirmed by molecular biology. Symptoms included the daily occurrence of blood-filled blisters and permanent itching. Wounds were easily infected and necrotic maroon in color. The patient was in severe pain and required frequent use of morphine. Did not respond during the first 6 months of life to topical treatment qualified by the EB international guidelines (moisturizers, antibiotic creams, and dressings.

Method/Intervention: Treatment of skin and oral mucosa with a topical ointment elaborated with petrolatum, cod liver oil, virgin beeswax, sunflower oil, sweet almond oil, and vitamins A, D, and E. Dressings changed every 12 hours: ointment on sterile gauzes covered by dressings to allow for oxygenation; no wound touching or cleaning. The ointment was spread on the lips, oral mucosa, and tongue.

Quality of Life (QoL) first 6 months of life pre-intervention: Pain as dominant symptom treated with opioids. Generalized serohematic blisters in the skin, oral mucosa, and lips; aplasia cutis with repeated infections in feet. Intense pruritus and scratching alter sleep. Frequent crying, feeding difficulty, and sleep alteration. Patient needs specialized nursing assistance every 2–4 hours. (**Figure 3a–e**, Video 1: https://youtu.be/8kiL-9hrJXU).

QoL after 6 months of life post-intervention: - Upon treatment onset: Significant pain relief and reduction of blister number and size, blood-free content, night itching calms down. Sleep regularizes, and feeding improves. After 20 days of treatment, opioids are discontinued. –.

At 6 months of age: cure of aplasia cutis. Cure wounds every 24 hours by the parents to date. The wounds never got infected (**Figure 3f, g**).

After 120 days of treatment with Curefini™ ointment (**Figure 3h, i**).

At 3 years of age, continuous positive skin response was observed (**Figure 3j**).

At 6 years of age: until then, baths with chlorhexidine in water, thereafter she showered independently, with wounded areas covered by dressings. Patient gets dressed with assistance.

Schooling: At age 3 she initiated pre-school, continuing to elementary school with good performance.

First esophageal dilation at the age of 5 years; then 5th and last at the age of 8 years in 2019. Off-label treatment with Losartan potassium since May 2020 (REFLECT), dose 17.5 mg b.i.d. Entertainment: Rides a tricycle, on horseback, and interacts with friends. Current hematological parameters: Anemia, HB levels never below 10 g/dL; HCT: 31.4%. ESR: 27 mm/hrs. (0–20); PCR: 15 mg/L (< 5 mg/L). Treatment: Iron poly maltose 1 mL/d (50 mg). Liver and kidney function: NAD.

At 9 years of age: BMI 14.2 (WHO child growth standards; p 3/15); body weight 23.700 kg; height: 123 cm. No pseudo syndactyly or microstomy, good mouth opening, dentition NAD. Persistent pruritus was informed. (**Figure 3k**).

Conclusion: During 9 years of observation and treatment, the ointment controlled pain, reduced affected areas, and prevented skin infections and retractile scars.

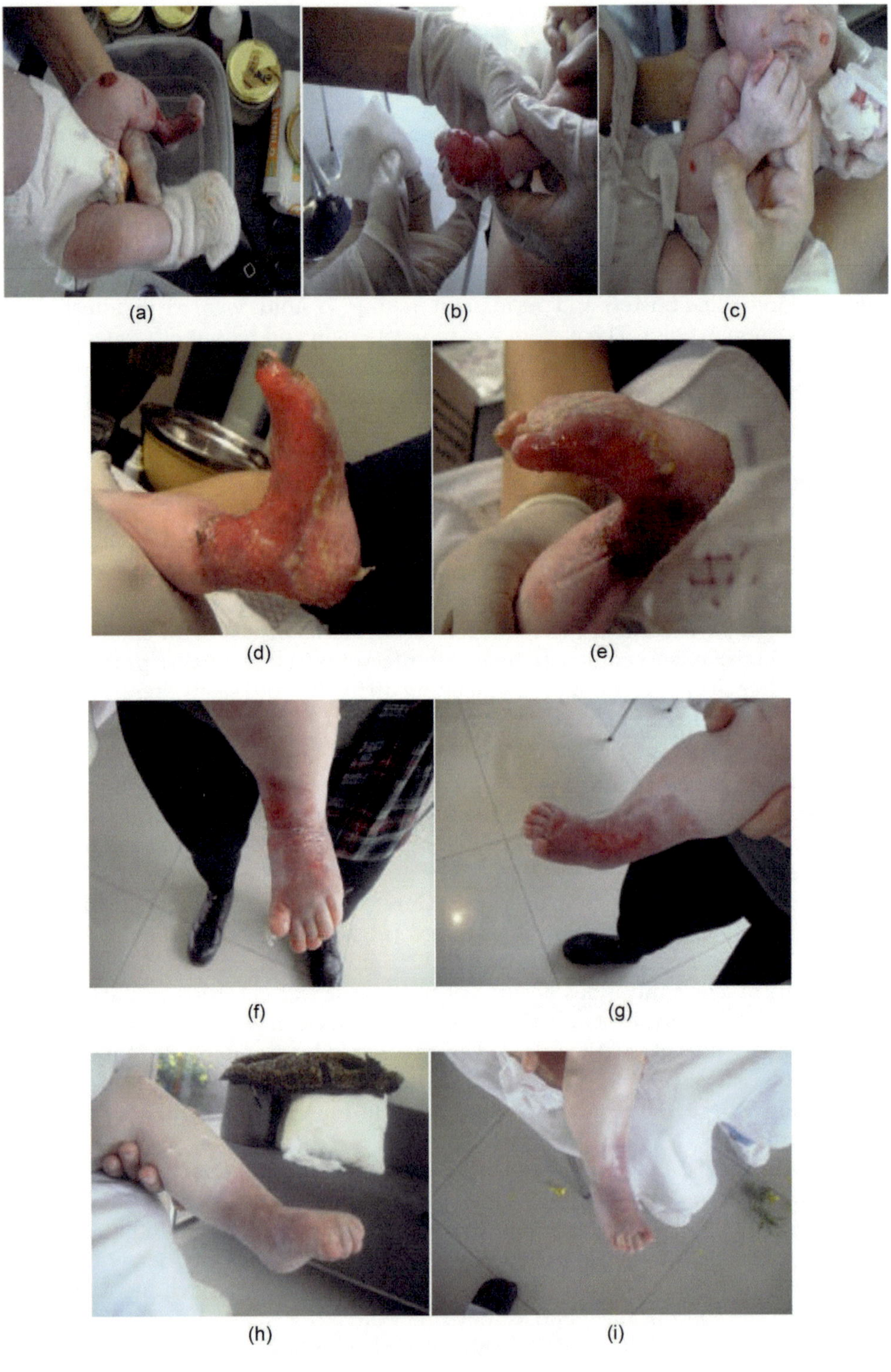
(a)
(b)
(c)
(d)
(e)
(f)
(g)
(h)
(i)

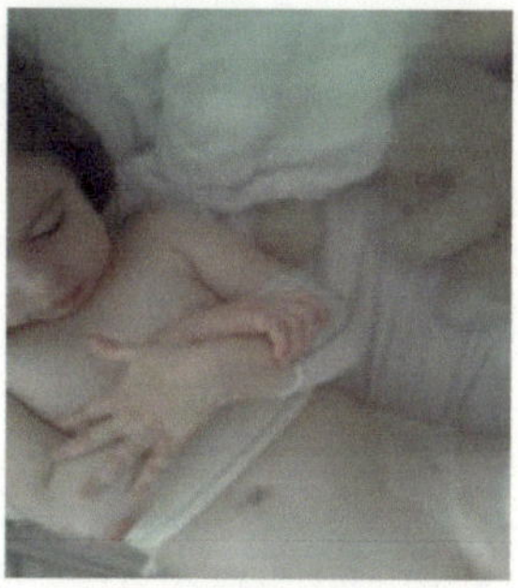

(j)

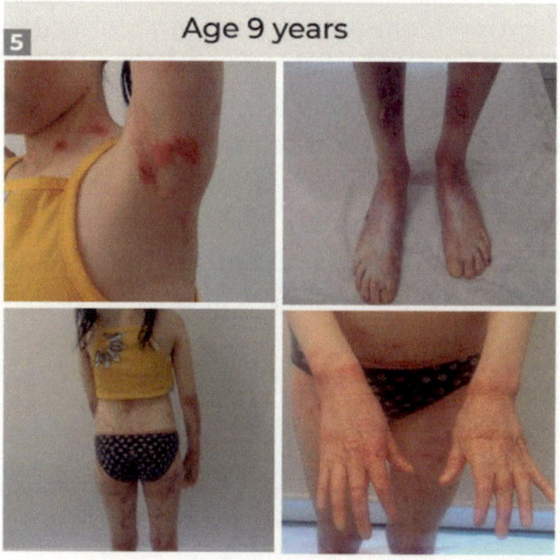

(k)

(l)

Figure 3.
(a–c) New-born conventional treatment. (d, e) 5 month old, conventional treatment. (f, g) 6 ½ month old, 20-days long treatment with Curefini™ ointment. (h, i) after 120 day of treatment with Curefini™ ointment. (j) after 3 years of treatment with Curefini™ ointment.

Well, patient compliance to treatment method. Well integrated into school and peers. CDLQI = 5 (**Figure 3l**, Video 2: https://youtu.be/i2_Myo2VnRY).

6. Conclusions

Natural source combined formula of cod liver oil, sweet almond oil, sunflower oil, and beeswax ointment has clinically proven its efficacy as an enhancer of the wound healing process showing particular efficacy on controlling skin inflammation, thus local pain. Nor clinical dermatitis or skin irritation was detected while it is used. It can be used safely both on skin and mucosa.

If proper sterile application technique is used, no bioburden or wound infection was detected as a complication in any of the three different clinical cases.

Beyond its own properties, the clinical use of this particular balanced natural source formula has reduced the concomitant use of topical and systemic antibiotics, corticosteroids and opioids.

Topical use of PUFA's and beeswax may lead the way for a new kind of medicine understanding that humans belong to Earth and we are here not only to take care of ourselves but to protect what can heal us.., Nature!

Conflict of interest

The authors declare no conflict of interest.

Author details

Irina Saretzky[1*] and Marta Cassini[2]

1 Wound Healing Section, Dermatology Department, Hospital de Clínicas José de San Martín, University of Buenos Aires, Argentina

2 Epidermolysis Bullosa Research, University of Buenos Aires, Argentina

*Address all correspondence to: irinasares@yahoo.com.ar

References

[1] Niculescu A-G, Grumezescu AM. An up-to-date review of biomaterials application in wound management. Polymers. 2022;**14**(3):421. DOI: 10.3390/polym14030421

[2] Stoica AE, Chirkov C, Grumezescu AM. Nanomaterials for wound dressings: An up-to-date overview. Molecules. 2020;**25**(11):2699. DOI: 10.3390/molecules25112699

[3] Brudzynski K. Honey as an ecological reservoir of antibacterial compounds produced by antagonistic microbial interactions in plant nectars, honey and honey bee. Antibiotics. 2021;**10**(5):551. DOI: 10.3390/antibiotics10050551

[4] Cornara L, Biagi M, Xiao J, Burlando B. Therapeutic properties of bioactive compounds from different honeybee products. Frontiers in Pharmacology. 2017;**8**:412

[5] Fratini F, Cilia G, Turchi B, Felicioli A. Beeswax: A minireview of its antimicrobial activity and its application in medicine. Asian Pacific Journal of Tropical Medicine. 2016;**9**(9):839-843

[6] Leiva-Sabadini C et al. Antibacterial effect of honey-derived exosomes containing antimicrobial peptides against Oral streptococci. International Journal of Nanomedicine. 2021;**16**:4891-4900. DOI: 10.2147/IJN.S315040

[7] Jara CP, Mendes NF, Prado T, de Araújo EP. Bioactive fatty acids in the resolution of chronic inflammation in skin wounds. Advances in Wound Care. 2020;**9**(8):472-490. DOI: 10.1089/wound.2019.1105

[8] Terkelsen LH, Eskild-Jensen A, Kjeldsen H, Barker JH, Hjortdal VE. Topical application of cod liver oil ointment accelerates wound healing: An experimental study in wounds in the ears of hairless mice. Scandinavian Journal of Plastic and Reconstructive Surgery and Hand Surgery. 2000;**34**(1):15-20

[9] Addisu MS. Effect of topical application of mixture of cod liver oil and honey on old (chronic) wounds and granulation tissue in donkeys. Journal of Veterinary Medicine and Animal Health. 2018;**10**(3):78-88

[10] Ahmed J, Bannai JA, Abdulsamad RK. Topical application of mixture of honey with cod liver oil as a novel therapy to treat contaminated skin wounds in rabbits model. Kufa Journal for Veterinary Medical Sciences. 2017;**8**(1):19-30

[11] Saretzky I. Second degree burn treated with Curefini ointment and PVC film: An accessible wound care strategy in developing countries. Wounds International. 2021;**12**(1):40-43

[12] Manzur G, Cassini M, Natale M, Valinotto L, Massimo I, Massimo H. Evaluation of topical therapy with the use of ointment made with natural ingredients in patients with recessive dystrophic epidermolysis bullosa. In: Proceedings of the DEBRA Congress International, Zermat Switzerland. 2018

[13] Wound Source Wound care 101. Available from: https://www.woundsource.com/blog/four-stages-wound-healing

[14] Wallace HA, Basehore BM, Zito PM. Wound Healing Phases. Treasure Island (FL): StatPearls Publishing; 2023

[15] Bowden LG, Byrne HM, Maini PK, Moulton DE. A morphoelastic model for

dermal wound closure. Biomechanics and Modeling in Mechanobiology. 2016;**15**(3):663-681

[16] Percival SL, McCarty SM, Lipsky B. Biofilms and wounds: An overview of the evidence. Advances in Wound Care. 2015;**4**:373-381

[17] Wolcott R. Dowd S the role of biofilms: Are we hitting the right target? Plastic and Reconstructive Surgery. 2011;**127**(Suppl 1):28S-35S

[18] Preda VG, Săndulescu O. Communication is the key: Biofilms, quorum sensing, formation and prevention. Discoveries (Craiova). 2019;7(3):e100. DOI: 10.15190/d.2019.13

[19] Ganesh K, Sinha M, Mathew-Steiner SS, Das A, Roy S. Sen CK chronic wound biofilm model. Advances in Wound Care. 2015;**4**:382-388

[20] Lebeaux D, Chauhan A, Rendueles O. Beloin C from in vitro to in vivo models of bacterial biofilm-related infections. Pathogens. 2013;**2**:288-356

[21] Roy S, Elgharably H, Sinha M, et al. Mixed-species biofilm compromises wound healing by disrupting epidermal barrier function. The Journal of Pathology. 2014;**233**:331-343

[22] Kim MH. Nanoparticle-based therapies for wound biofilm infection: Opportunities and challenges. IEEE Transactions on Nanobioscience. 2016;**15**:294-304

[23] Gilmore BF, Flynn PB, O'Brien S, Hickok N, Freeman T, Bourke P. Cold plasmas for biofilm control: Opportunities and challenges. Trends in Biotechnology. 2018;**36**:627-638

[24] Leaper D. Appropriate use of silver dressings in wounds: International consensus document. International Wound Journal. 2012;**9**:461-464

[25] Chen X. Schluesener HJ Nanosilver: A nanoproduct in medical application. Toxicology Letters. 2008;**176**:1-12

[26] Fitzgerald DJ, Renick PJ, Forrest EC, et al. Cadexomer iodine provides superior efficacy against bacterial wound biofilms in vitro and in vivo. Wound repair and regeneration: official publication of the Wound Healing Society [and] the European Tissue Repair Society. 2017;**25**:13-24

[27] Romanowski EG, Stella NA, Yates KA, Brothers KM, Kowalski RP. Shanks RMQ In vitro evaluation of a Hypochlorous acid hygiene solution on established biofilms. Eye & Contact Lens. 2018;**44**(Suppl 2):S187-S191

[28] Day A, Alkhalil A, Carney BC, Hoffman HN, Moffatt LT, Shupp JW. Disruption of biofilms and neutralization of bacteria using Hypochlorous acid solution: An In vivo and In vitro evaluation. Advances in skin and wound care. 2017;**30**:543-551

[29] Papenfort K. Bassler BL quorum sensing signal-response systems in gram-negative bacteria. Nature Reviews. Microbiology. 2016;**14**:576-588

[30] Lee JH, Park JH, Kim JA, et al. Low concentrations of honey reduce biofilm formation, quorum sensing, and virulence in Escherichia coli O157: H7. Biofouling. 2011;**27**:1095-1104

[31] He Z, Wang Q, Hu Y, et al. Use of the quorum sensing inhibitor furanone C-30 to interfere with biofilm formation by Streptococcus mutans and its luxS mutant strain. International Journal of Antimicrobial Agents. 2012;**40**:30-35

[32] Endo A, Futagawa-Endo Y, Dicks LM. Isolation and characterization of

fructophilic lactic acid bacteria from fructose-rich niches. Systematic and Applied Microbiology. 2009;**32**:593-600

[33] Endo A, Salminen S. Honeybees and beehives are rich sources for fructophilic lactic acid bacteria. Systematic and Applied Microbiology. 2013;**36**:444-448

[34] Berríos P, Fuentes JA, Salas D, Carreño A, Aldea P, Fernández F, et al. Inhibitory effect of biofilm-forming lactobacillus kunkeei strains against virulent Pseudomonas aeruginosa in vitro and in honeycomb moth (galleria mellonella) infection model. Beneficial Microbes. 2018;**9**:257-268

[35] Aween MM, Hassan Z, Muhialdin BJ, Noor HM, Eljamel YA. Evaluation on antibacterial activity of lactobacillus acidophilus strains isolated from honey. American Journal of Applied Sciences. 2012;**9**:807-817

[36] Muhialdin BJ, Hassan Z, Saari N. In vitro antifungal activity of lactic acid bacteria low molecular peptides against spoilage fungi of bakery products. Annales de Microbiologie. 2018;**68**:557-567

[37] Sadiq FA, Yan B, Tian F, Zhao J, Zhang H, Chen W. Lactic acid bacteria as antifungal and anti-Mycotoxigenic agents: A comprehensive review. Comprehensive Reviews in Food Science and Food Safety. 2019;**18**:1403-1436

[38] Jack RW, Tagg JR, Ray B. Bacteriocins of gram-positive bacteria. Microbiological Reviews. 1995;**59**:171

[39] Caulier S, Nannan C, Gillis A, Licciardi F, Bragard C, Mahillon J. Overview of the antimicrobial compounds produced by members of the Bacillus subtilis group. Frontiers in Microbiology. 2019;**10**:302

[40] Pajor M, Worobo RW, Milewski S, Szweda P. The antimicrobial potential of bacteria isolated from honey samples produced in the apiaries located in Pomeranian Voivodeship in northern Poland. International Journal of Environmental Research and Public Health. 2002;**2018**:15

[41] Grecka K, Kus PM, Worobo RW, Szweda P. Study of the anti-staphylococcal potential of honeys produced in northern Poland. Molecules. 2018;**23**:26

[42] Snowdon JA, Cliver DO. Microorganisms in honey. International Journal of Food Microbiology. 1996;**31**:1-26

[43] Alvarez-Pérez S, Herrera CM, de Vega C. Zooming-in on floral nectar: A first exploration of nectar-associated bacteria in wild plant communities. FEMS Microbiology Ecology. 2012;**80**:591-602

[44] Rodríguez-Andrade E, Stchigel AM, Terrab AJ, Guarro AJ, Cano-Lira JF. Diversity of xerotolerant and xerophilic fungi in honey. IMA Fungus. 2019;**10**:20

[45] Breukink E, Wiedemann I, van Kraaij C, Kuipers OP, Sahl HG, de Kruijff B. Use of the cell wall precursor lipid II by a pore-forming peptide antibiotic. Science. 1999;**286**:2361-2364

[46] Moll GN, Konings WN, Driessen AJ. Bacteriocins: Mechanism of membrane insertion and pore formation. Antonie Van Leeuwenhoek. 1999;**76**:185-198

[47] National Pressure Ulcer Advisory Panel (NPUAP) announces a change in terminology from pressure ulcer to pressure injury and updates the stages of pressure injury|the National Pressure Ulcer Advisory Panel—NPUAP, Available from: https://www.npuap.

org/national-pressure-ulcer-advisory-panel-npuap-announcesa-change-in-terminology-from-pressure-ulcer-to-pressureinjury-and-updates-the-stages-of-pressure-injury/.

[48] Meddings J, Reichert H, Rogers MAM, Hofer TP, McMahon LF Jr, Grazier KL. Under pressure: Financial effect of the hospital-acquired conditions initiative-a statewide analysis of pressure ulcer development and payment. Journal of American Geriatrics Society. 2015;**63**(7):1407-1412

[49] Zuñiga-Castro C, Espinoza-Caceres NA, Fernandez-Pacheco MA. Level of risk and incidents of pressure ulcers in hospitalized patients in the medical services of a national hospital. Revista Enfermería Herediana. 2015;7(2):155

[50] Prieto JMR, Areosa LA, Romero M, Romero M, Collado FG, Pizano AR. Adverse events related to pressure ulcers: An avoidable problem? Enfermería Dermatológica. 2010;**4**(11):21-26

[51] Baumgarten M, Margolis DJ, Localio AR, et al. Pressure ulcers among elderly patients early in the hospital stay. Journals of Gerontology Series A: Biological Sciences and Medical Sciences. 2006;**61**(7):749-754

[52] VanGilder C, MacFarlane G, Meyer S, Lachenbruch C. Body mass index, weight, and pressure ulcer prevalence: An analysis of the 2006-2007 international pressure ulcer prevalence surveys. Journal of Nursing Care Quality. 2009;**24**(2):127-135

[53] Drake DJ, Swanson M, Baker G, et al. The association of BMI and Braden total score on the occurrence of pressure ulcers. The Journal of Wound, Ostomy and Continence Nursing. 2010;**37**(4):367-371

[54] Bergstrom N, Braden B, Kemp M, Champagne M, Ruby E. Multi-site study of incidence of pressure ulcers and the relationship between risk level, demographic characteristics, diagnoses, and prescription of preventive interventions. Journal of the American Geriatrics Society. 1996;**44**(1):22-30

[55] Bergstrom N, Braden B, Kemp M, Champagne M, Ruby E. Predicting pressure ulcer risk: A multisite study of the predictive validity of the Braden scale. Nursing Research. 1998;**47**(5):261-269

[56] Charlson ME, Pompei P, Ales KL, Mackenzie CR. A new method of classifying prognostic comorbidity in longitudinal studies: Development and validation. Journal of Chronic Diseases. 1987;**40**(5):373-383

[57] Flattau A, Blank AE. Risk factors for 90-day and 180- day mortality in hospitalized patients with pressure ulcers. International Wound Journal. 2014;**11**(1):14-20

[58] Averill RF, Goldfield N, Hughes JS, et al. Development of the all Patient Refined DRGs (APR-DRGs) Research Report, 3M Health Information Systems, Wallingford, CT. 2003. Available from: www.3Mhis.com

[59] Tschannen D, Bates O, Talsma A, Guo Y. Patientspecific and surgical characteristics in the development of pressure ulcers. American Journal of Critical Care. 2012;**21**(2):116-125

[60] Lumbley JL, Ali SA, Tchokouani LS. Retrospective review of predisposing factors for intraoperative pressure ulcer development. Journal of Clinical Anesthesia. 2014;**26**(5):368-374

[61] Agarwal VK, Kumar D, Kansal S, Khanna C. Negative pressure wound therapy versus conventional wound

therapy in pressure sores. International Journal of Science and Research (IJSR). 2017;**6**(3):794-798

[62] Batra RK, Aseeja V. VAC therapy in large infected sacral pressure ulcer grade IV-can be an alternative to flap reconstruction. Indian Journal of Surgery. 2014;**76**(2):162-164

[63] Hu FX, Hu XX, Yang XL, et al. Treatment of large avulsion injury in perianal, sacral, and perineal regions by island flaps or skin graft combined with vacuum assisted closure. BMC Surgery. 2019;**19**(1):65

[64] Lee KY, Mooney DJ. Alginate: Properties and biomedical applications. Progress in Polymer Science. 2012;**37**(1):106-126

[65] Goertz O, Abels C, Knie U, May T, Hirsch T, Daigeler A, et al. Clinical safety and efficacy of a novel thermoreversible polyhexanide-preserved wound covering gel. European Surgical Research. 2010;**44**(2):96-101

[66] Matica MA, Aachmann FL, Tøndervik A, Sletta H, Ostafe V. Chitosan as a wound dressing starting material: Antimicrobial properties and mode of action 1. International Journal of Molecular Sciences. 2019;**20**:5889. DOI: 10.3390/ijms20235889

[67] Molan P, Rhodes T. Honey: A biologic wound dressing. Wounds. 2015;**27**(6):141-151

[68] Powell G. Evaluating Cutimed Sorbact: Using a case study approach. The British Journal of Nursing. 2009;**18**(15):S30

[69] Available from: https://www.wounds-uk.com/resources/details/suprasorb-xphmb-made-easy---products-for-practice

[70] Benskin LL. Poly mem((R)) Wic((R)) silver((R)) rope: A multifunctional dressing for decreasing pain, swelling, and inflammation. Adv Wound Care (New Rochelle). 2012;**1**(1):44-47

[71] World Health Organization Burns Fact Sheets. Available from: https://www.who.int/news-room/fact-sheets/detail/burns

[72] Observatorio de Femicidios. Defensoría del Pueblo de la Nación Argentina. Available from: https://www.dpn.gob.ar/observatorio-femicidios.php

[73] Marques SR, Peixoto CA, Messias JB, Albuquerque AR, Silva VA. The effects of topical application of sunflower-seed oil on open wound healing in lambs. Acta Cirurgica Brasileira. 2004;**19**(3):196-209. DOI: 10.1590/s0102-86502004000300005

[74] Hajhashemi M, Rafieian M, Rouhi Boroujeni HA, et al. The effect of aloe vera gel and sweet almond oil on striae gravidarum in nulliparous women. Journal of Maternal-Fetal and Neonatal Medicine. 2018;**31**(13):1703-1708. DOI: 10.1080/14767058.2017.1325865

[75] Davis SC, Lie J, Gil J, Vades J, Solis M. Study Report Second Degree Burn Porcine Model with Curefini Vs. Petrolatum Vehicle and Non Adherent Gauze. Coral Gables, Fl, USA: University of Miami, Miller School of Medicine. Departament of Dermatology and Cutaneous Surgery; 2014

[76] Nelson DL, Cox MM. Lehninger Principles of Biochemistry. New York: Worth Publishers; 2000

[77] Chapkin RS, Ziboh VA. Inability of skin enzyme preparations to biosynthesize arachidonic acid from linoleic acid. Biochemical and Biophysical Research Communications. 1984;**124**:784-792

[78] Calder PC. Omega-3 fatty acids and inflammatory processes: From molecules to man. Biochemical Society Transactions. 2017;**45**:1105-1115

[79] Kendall AC, Nicolaou A. Bioactive lipid mediators in skin inflammation and immunity. Progress in Lipid Research. 2013;**52**:141-164

[80] Dennis EA, Norris PC. Eicosanoid storm in infection and inflammation. Nature Reviews. Immunology. 2015;**15**:511-523

[81] Harris SG, Padilla J, Koumas L, et al. Prostaglandins as modulators of immunity. Trends in Immunology. 2002;**23**:144-150

[82] Nicolaou A. Eicosanoids in skin inflammation. Prostaglandins, Leukotrienes, and Essential Fatty Acids. 2013;**88**:131-138

[83] Zhao J, Shu B, Chen L, et al. Prostaglandin E2 inhibits collagen synthesis in dermal fibroblasts and prevents hypertrophic scar formation in vivo. Experimental Dermatology. 2016;**25**:604-610

[84] Sato T, Kirimura Y, Mori Y. The co-culture of dermal fibroblasts with human epidermal keratinocytes induces increased prostaglandin E2 production and cyclooxygenase 2 activity in fibroblasts. The Journal of Investigative Dermatology. 1997;**109**:334-339

[85] Silva JC, Pitta MGR, Pitta IR, et al. New peroxisome proliferator-activated receptor agonist (GQ-11) improves wound healing in diabetic mice. Advanced Wound Care (New Rochelle). 2019;**8**:417-428

[86] Mirza RE, Fang MM, Novak ML, et al. Macrophage PPARgamma and impaired wound healing in type 2 diabetes. The Journal of Pathology. 2015;**236**:433-444

[87] Navarro-Hortal MD, Orantes-Bermejo FJ, Sánchez-González C, Varela-López A, Giampieri F, Torres F-PC, et al. Industrial-scale decontamination procedure effects on the content of Acaricides, heavy metals and antioxidant capacity of beeswax. Molecules. 2019;**24**:1518. DOI: 10.3390/molecules24081518

[88] Karabey T, Karag ÖZoĞLu Ş. Using beeswax, olive oil and centaury oil for pressure ulcers. Turkiye Klinikleri Journal of Medical Sciences. 2019;**27**:153-156. DOI: 10.5336/caserep.2019-64740

[89] Zhang Y, Simpson BK, Dumont M-J. Effect of beeswax and carnauba wax addition on properties of gelatin films: A comparative study. Food Bioscience. 2018;**26**:88-95. DOI: 10.1016/j.fbio.2018.09.011

[90] Nader RA, Mackieh R, Wehbe R, El Obeid D, Sabatier JM, Fajloun Z. Beehive products as antibacterial agents: A review. Antibiotics. 2021;**10**:717. DOI: 10.3390/antibiotics10060717

[91] Cassini M, Saretzky I. Grade IV sacral sore treated with an ointment rich in PUFAs, ceramides, and antimicrobial peptides. Case Reports in Dermatological Medicine. 2022;**2022**:4445055, 8 pages. DOI: 10.1155/2022/4445055

[92] National Pressure Ulcer Advisory Panel. European pressure ulcer advisory panel and pan pacific pressure injury alliance. In: Emily H, editor. Prevention and Treatment of Pressure Ulcers: Clinical Practice Guideline. Perth, Australia: Cambridge Media; 2014

[93] Toriyabe S, Saito H, Sakurai K. Short Reports: Use of a food as a dressing material. Advances in Wound Care. Oct 1999;**12**(8):405-406

Chapter 5

Thermal Behavior of Waxes and Its Correlation with Mascara Stability Tests: A DSC Study

Ricardo Alejandro Pineda Beltrán and Johnbrynner García

Abstract

In this work, eight commercial eyelashes mascaras were stored during 15 days at 50°C and a correlation between visually perceived separations over time and the thermal behavior of the waxes present within each product was observed. It was found that 3 of the 8 studied mascaras, displayed noticeable separations which agreed with both changes in their DSC curves and the presence of 2 heat-sensitive waxes in their formula. Finally, it is suggested that the addition of heat-sensitive wax mixtures into a formula should be discarded in order to avoid visual phase-separations and produce thermally stable mascaras that offer the product experience designed for users.

Keywords: cosmetics, differential scanning calorimetry, heat-sensitive materials, mascaras, thermal stability, waxes

1. Introduction

Despite waxes are among the oldest materials known by humans, the word wax covers a wide number of compounds and there is not a generally accepted definition nowadays for that term. According to Krendlinger et al., waxes are substances that are kneadable and not glassy at 20°C, have a drop point >40°C, their melt viscosity does not exceed the 10,000 mPa s at 10°C above the drop point, and usually melt between 50 and 90°C without decomposition [1].

Waxes are extensively used in ceramics, adsorbents, adhesives, paints, medicines and mainly in cosmetics, defining binding and surface properties [1]. Therefore, having a characterization technique for those materials is necessary.

Due to their thermal properties, with the aim of differentiate and identify waxes, Differential Scanning Calorimetry (DSC) is preferred over physical properties or techniques as gas chromatography or infrared spectroscopy [2, 3]. DSC has been also employed in the quantification of wax blends [4] and the thermal stability evaluation after repeated heating cycles [5, 6].

Several studies have found that oil viscosities and wax ratios have great influence on the sensory and mechanical characteristics of cosmetic formulations [7, 8]. Among the skin cosmetics, the ones with a major amount of waxes are lipsticks [9]. Recently, commercially available benchmark lipstick prototypes were characterized using DSC and prototype-dependent fingerprinting was evaluated, mainly in terms of heat stability as well as spreading [10].

Mascaras are products used by 59% of female internet consumers (18+), that make eyelashes appear thicker, longer, darker and with an enhanced curl [11]. Using different techniques among which DSC is reported [12], mascaras are one of the most studied cosmetics since they have some of the highest consumer expectations to meet [11]. As used in lipsticks, a mascara formula contains wax blends to achieve the desired texture [13]; however, to our knowledge, there is not a published research linking the thermal properties of waxes and a key attribute for consumers as the perceived visual appearance [14] (affected by the thermal stability of the product).

The aim of this chapter was to study the relation between the thermal behavior of waxes by means of DSC and the thermal stability screening test of eight commercial mascaras showing the advantages of the continuous development of calorimetry applications in the cosmetic industry focused in efficacy and quality.

2. Materials and methods

2.1 Sample source

The eight commercial eyelashes mascaras were used only for research purposes. Due to the company's policies, the name of each product, the proportions of the individual ingredients in the formula as well as the supplier for each wax are outside the scope of the current study.

The studied waxes were Carnauba wax (Wax-1), Microcrystalline wax (Wax-2), Ozokerite (Wax-3), Paraffin wax (Wax-4), Polyethylene wax (Wax-5) and Synthetic Beeswax (Wax-6).

2.2 Thermal stability screening test

Following a preliminary stability test (also known as screening test), mascara samples were heated at 50°C for 15 days to accelerate the instability processes that could occur, obtaining preliminary information on stable formulations [15]. After 0, 8 and 15 days of thermal stress, each product was carefully observed for phase-separation recording and enough amount of each sample was used to perform DSC measurements. Control samples (day 0) were measured as received from the supplier.

2.3 Differential scanning calorimetry measurements

The DSC curves were obtained using a NETZSCH DSC 204 F1 Phoenix differential scanning calorimeter (NETZSCH-Geratebau GmbH, Selb, Germany) calibrated with Indium, Tin and Bismuth. For each measurement an empty aluminum pan—previously weighted—was used as reference. Mascara and wax samples were weighed (5.5 ± 0.1 mg) and sealed in a 25 μl-aluminum pan and before each analysis, a small hole was opened in the top of the lid. The results were processed using the Proteus software version 7.2.0. The melting temperature (°C), and the transition enthalpy (ΔH, J/g) were estimated.

For waxes a constant rate of 10°C/min in the temperature range of −60 to 180°C was used and three successive heating cycles were recorded under dynamic N_2 atmosphere (20 mL/min) ensuring reliable and reproducible data [10].

Regarding the mascara samples, two heating cycles were used: the first from −20 to 120°C at 10°C/min, and from 0 to 120°C at 5°C/min for the second cycle [16]. All these experiments were made at least by duplicate.

3. Results and discussion

3.1 Wax characteristics

The curves reported on **Figure 1** allow to see the thermal behavior of 6 waxes present within the composition of the mascaras.

DSC curves of waxes 1, 2, 3 and 4 remained unchanged upon followed heating, which can be a stability indicator [17]. On the other hand, DSC curves of waxes 5 (polyethylene wax) and 6 (beeswax) did change.

Although beeswax samples can present DSC differences depending on the origin of the wax [18], they have a single endothermic peak with melting point above 50°C [19, 20] and there are no modifications in the DSC curve after 50 heating cycles [21]. However, the

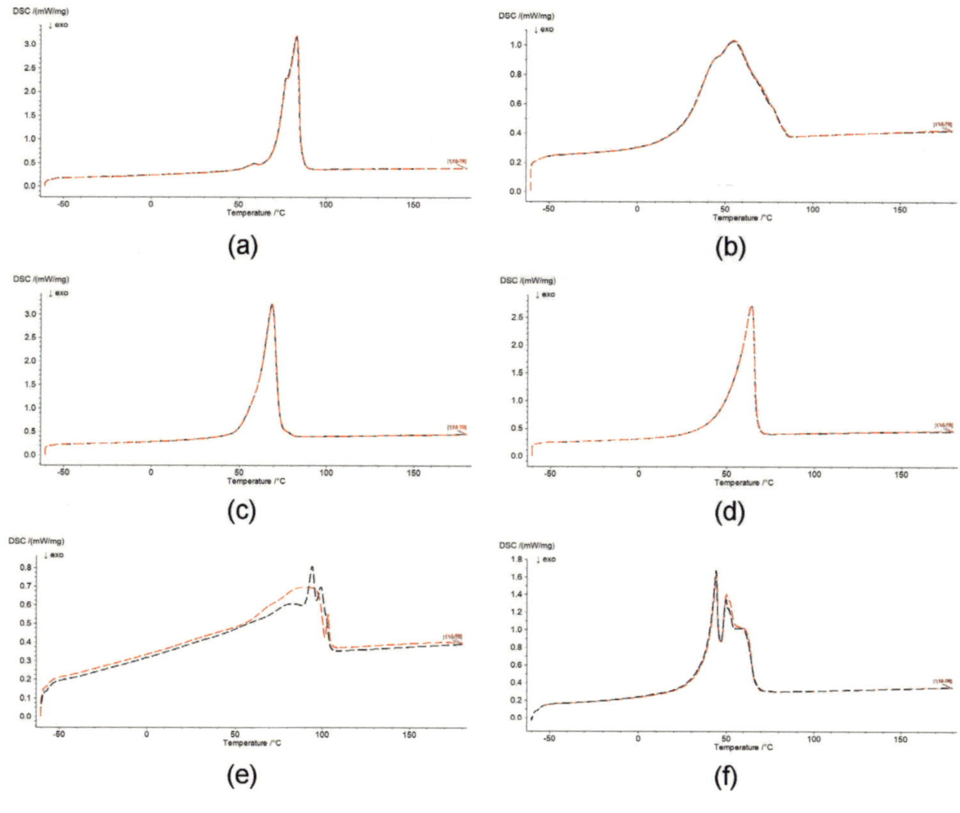

Figure 1.
DSC curves of waxes: Wax-1 (a), Wax-2 (b), Wax-3 (c), Wax-4 (d), Wax-5 (e), Wax-6 (f). In each case the red line (–) and the black line (–) correspond to different heating cycles.

so-called synthetic beeswax (wax-6) sample herein reported is different, since it has two melting endothermic peaks below 50°C and its DSC curve changes between heating cycles.

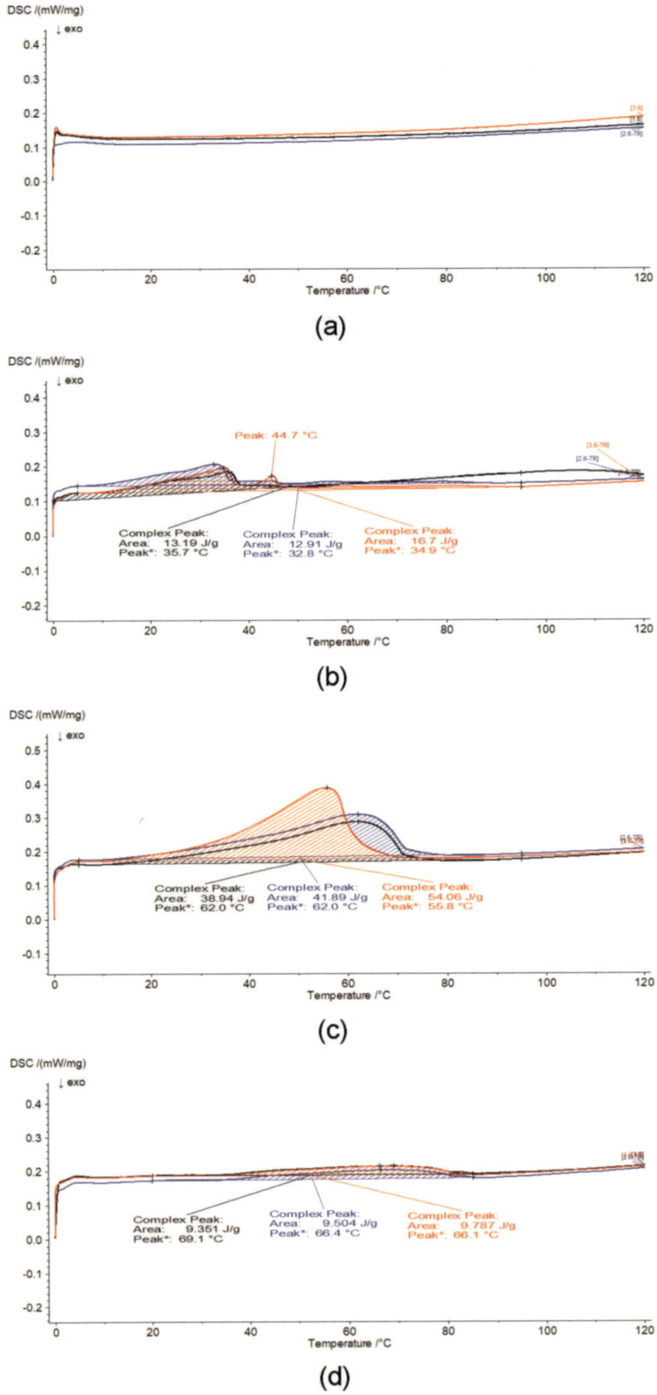

Figure 2
DSC curves of the mascaras M-1 (a), M-2 (b), M-3 (c), M-4 (d). *In each case the black line (–) corresponds to day 0, the blue line (–) to 8 days and the red line (–) to 15 days.*

A change in the thermal behavior of waxes after subsequent remelting cycles is unusual and has been taken as an indicator of variations in the wax structure [22]. For that reason, waxes 5 and 6 were considered as heat-sensitive.

3.2 Correlation with mascaras

Figures 2 and **3** shows the DSC curves for each mascara, while **Table 1** shows the composition of each one in terms of waxes.

Curves without signals (**Figures 2** and **3**) belong to the mascaras that did not have waxes in their composition M-1 and M-8, whereas the mascaras M-2, M-3, M-4, M-5, M-6 and M-7 did show DSC signals in the range between 5 and 95°C and were made of waxes. Also, since mascaras are made of water, pigments, waxes, film formers, emulsifiers, humectants, viscosity agents, preservatives and fibers [11] and waxes

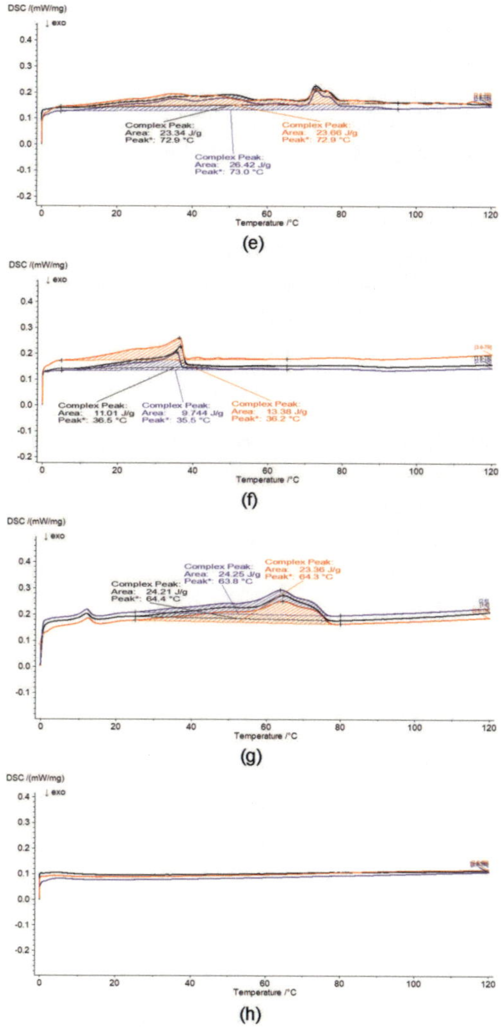

Figure 3.
DSC curves of the mascaras M-5 (e), M-6 (f), M-7 (g) and M-8 (h). In each case the black line (–) corresponds to day 0, the blue line (–) to 8 days and the red line (–) to 15 days.

Wax	M-1	M-2	M-3	M-4	M-5	M-6	M-7	M-8	Heat-sensitive?
Wax-1							X		No
Wax-2			X	X					No
Wax-3			X						No
Wax-4					X		X		No
Wax-5		X	X	X	X	X			Yes
Wax-6		X	X			X	X		Yes

Table 1.
Waxes present in each mascara and their change in DSC curves between heating cycles.

have a melting range between 8 and 100°C [2], the DSC signals are produced by the waxes within the mascara composition.

Although Wax-5 and Wax-6 were labeled as heat-sensitive waxes, mascaras containing one or another (M-4, M-5 and M-7) did not undergo significant changes in their thermal properties in function of time. Likewise, according to **Table 2** (which differentiates the mascaras that presented a noticeable phase-separation from those that did not after being subjected to 50°C for 15 days) M-4, M-5 and M-7 did not show visual separation.

Due to signals on the mascara DSC curve are generated by waxes, it was hypothesized that alterations in the curves of some mascaras could be a consequence of physicochemical changes in the constituent waxes, which is in agreement with evident alterations for mascaras which contained the blend of both heat sensitive waxes (M-2, M-3 and M-6). When comparing the stress at 50°C for 15 days with the other two control times (0 and 8 days at 50°C), an increase in peak enthalpies in more than 20% and separation phenomena were observed. Then, it is possible to say that the mixture of heat-sensitive waxes is one source of incompatibility that yields unstable mascaras.

Given that, it is recommended to check whether the waxes used for mascara manufacture are heat-sensitive. It is necessary to study the specific physicochemical interaction of waxes as well as their ratios, in order to continue the development of meaningful information to predict the stability of mascaras.

Mascara code	Visual changes?	Peak enthalpy variation (%)*
M-1	No	N.A
M-2	Yes	26.6
M-3	Yes	38.8
M-4	No	4.7
M-5	No	1.4
M-6	Yes	21.5
M-7	No	−3.5
M-8	No	N.A

**Respect to day 0.*

Table 2.
Peak enthalpy variation and visual changes after 15 days at 50°C of the analyzed mascaras.

4. Conclusions

Changes in DSC curves of Wax-5 and Wax-6 when applying successive heating cycles were observed, indicating their heat-sensitive nature. Signals between 5 and 95°C on the DSC curve of the studied mascaras are related with the presence of waxes in their composition. Mascaras M-2, M-3 and M-6 displayed changes in their curves after 15 days at 50°C that were related with noticeable phase-separation, such separation could be a consequence of destabilization phenomena after blending at least two heat-sensitive waxes.

Acknowledgements

The authors acknowledge to all the Laboratory team from the Research, Development & Innovation Area of Belcorp for the help performing some measurements, getting all the mascara compositions and the fruitful academic discussions.

Conflicts of interest

The authors declare that they have no conflict of interest.

Funding source

The current work was supported by Belcorp Company.

Author details

Ricardo Alejandro Pineda Beltrán and Johnbrynner García*
Belcorp, Research, Development and Innovation, Tocancipá, Colombia

*Address all correspondence to: johngarcia@belcorp.biz

References

[1] Krendlinger E, Wolfmeier U, Schmidt H, Heinrichs F-L, Michalczyk G, Payer W, et al. Waxes. In: Ullmann's Encycl. Ind. Chem. American Cancer Society; 2015. pp. 1-63. DOI: 10.1002/14356007.a28_103.pub2

[2] Flaherty B. Characterisation of waxes by differential scanning calorimetry. Journal of Applied Chemistry and Biotechnology. 1971;**21**:144-148. DOI: 10.1002/jctb.5020210507

[3] Miller R, Dawson G. Characterization of hydrocarbon waxes and polyethylenes by DSC. Thermochimica Acta. 1980;**41**:93-105. DOI: 10.1016/0040-6031(80)80098-0. Available from: https://www.sciencedirect.com/science/article/abs/pii/0040603180800980

[4] Stank J, Mullay J. Analysis of wax/oil mixtures using DSC. Thermochimica Acta. 1986;**105**:9-17. DOI: 10.1016/0040-6031(86)85218-2. Available from: https://www.sciencedirect.com/science/article/abs/pii/0040603186852182

[5] Al-Sammerrai D. Evaluation of the thermal stabilities of paraffin and microcrystalline wax melts. Thermochimica Acta. 1985;**86**:369-373. DOI: 10.1016/0040-6031(85)87067-2. Available from: https://www.sciencedirect.com/science/article/abs/pii/0040603185870672

[6] Kotsiomiti E, McCabe JF. Stability of dental waxes following repeated heatings. Journal of Oral Rehabilitation. 1995;**22**:135-143. DOI: 10.1111/j.1365-2842.1995.tb00247.x

[7] Clermont-Gallerande D et al. Relations between the sensory properties and fat ingredients of lipsticks. OCL. 2018;**25**:D502. DOI: 10.1051/ocl/2018053. Available from: https://www.ocl-journal.org/articles/ocl/full_html/2018/05/ocl180017/ocl180017.html

[8] De Clermont-Gallerande H, Chavardes V, Zastrow L. Crystal clear transparent lipstick formulation based on solidified oils. International Journal of Cosmetic Science. 1999;**21**:413-424. DOI: 10.1046/j.1467-2494.1999.211909.x

[9] Schneider G, Gohla S, Schreiber J, Kaden W, Schönrock U, Schmidt-Lewerkühne H, et al. Skin cosmetics. In: Ullmann's Encycl. Ind. Chem. American Cancer Society; 2001. DOI: 10.1002/14356007.a24_219

[10] Pan S, Germann N. Thermal and mechanical properties of industrial benchmark lipstick prototypes. Thermochimica Acta. 2019;**679**:178332. DOI: 10.1016/j.tca.2019.178332

[11] De Roeck A. Elevating mascaras: Formulating must- have lash effects with the Flick of a wand oil /wax phase. Cosmetics & Toiletries. 2019;**134**:1-4

[12] Ferrari V, Alfort P, et al. Method of making a mascara composition comprising a polyamide polymer and at least one inert filler. 2004

[13] Barel AO, Paye M, Maibach HI. Handbook of cosmetic science and technology. Dermatologic Surgery. 2010;**36**:664-667. DOI: 10.1111/j.1524-4725.2009.01444.x

[14] Crilly N, Moultrie J, Clarkson PJ. Seeing things: Consumer response to the visual domain in product design. Design Studies. 2004;**25**:547-577. DOI: 10.1016/j.destud.2004.03.001

[15] Baertschi SW, Hetrick EM, Hoaglund Hyzer CS, Pack BW, Roberts JC,

Wolfe CN. Chapter 11—Implementing an accelerated predictive stability program. In: Qiu F, Scrivens G, editors. Accel. Predict. Stab. Boston: Academic Press; 2018. pp. 253-283. DOI: 10.1016/B978-0-12-802786-8.00011-5

[16] Narebski S. Mascara composition comprising an ethylene/vinyl acetate copolymer, US8211416B2. 2012

[17] Pandey RK, Singh AK, Prakash R. Directed self-assembly of poly(3,3‴-dialkylquarterthiophene) polymer thin film: Effect of annealing temperature. Journal of Physical Chemistry C. 2014;**118**:22943-22951. DOI: 10.1021/jp507321z

[18] Buchwald R, Breed MD, Greenberg AR. The thermal properties of beeswaxes: Unexpected findings. Journal of Experimental Biology. 2008;**211**:121. DOI: 10.1242/jeb.007583

[19] Ihsan N, Teo Y, Najeeb Nellikkal M, Ahmad Z, Hasan A, Shah Z. Synthesis and characterization of magnetic nanostructured lipid carriers (mNLCs) for drug delivery. International Journal of Electrochemical Science. 2018;**13**:12045-12046. DOI: 10.20964/2018.12.02

[20] Galvão JG, Trindade GGG, Santos AJ, Santos RL, Chaves Filho AB, Lira AAM, et al. Effect of Ouratea sp. butter in the crystallinity of solid lipids used in nanostructured lipid carriers (NLCs). Journal of Thermal Analytical Calorimetry. 2016;**123**:941-948. DOI: 10.1007/s10973-015-4890-8

[21] Dinker A, Agarwal M, Agarwal GD. Experimental assessment on thermal storage performance of beeswax in a helical tube embedded storage unit. Applied Thermal Engineering. 2017;**111**:358-368. DOI: 10.1016/j.applthermaleng.2016.09.128

[22] Grzeskowiak K, Czarnecka-Komorowska D. Influence of waxes remelting used in investment casting on their thermal properties and linear shrinkage. Metalurgija. 2015;**54**:350-352

Chapter 6

A Novel Method for Titanium Dioxide Quantification in Cosmetic Products via Borate Fusion by Flame Atomic Absorption Spectroscopy

Cristian Rosales and Johnbrynner García

Abstract

A new method for extraction of titanium dioxide (TiO_2) from cosmetic matrices using borate salts for its quantification by flame atomic absorption spectroscopy (FAAS) was developed and validated. Following International Commission for Harmonization (ICH) and the United States Pharmacopeial Convention (USP) guidelines, the parameters of the method considered in this study were specificity, linearity, sensitivity, precision, and accuracy. In addition, critical factors of the method were assessed using a Youden–Steiner model. The method was able to differentiate the titanium signal from matrix and background signals, for which it is considered specific. The method is also linear for all cosmetic matrices and the raw material in the range 20–80 ppm with LOD and LOQ around 2 ppm and 7 ppm, respectively. Repeatability and intermediate precision were below 5.0%RSD, and Global Reproducibility was below 8.0% RSD. A digestion step free of HF or strong oxidizers makes this method a safer and easily transferable alternative to classical methods for quality control. It is thus a convenient tool for routine analysis of cosmetic products that need to comply with current regulations to ensure the safety of consumers and to guarantee batch-to-batch quality.

Keywords: cosmetics, TiO_2, FAAS, quality control, ultraviolet filter

1. Introduction

Skincare and makeup products are highly innovative and profitable segments of the cosmetic industry as they represent, together, near 60% of the total market size, which was estimated at around € 200B for 2019 [1]. Current consumer trends show an increased demand for products with a wide variety of functions but, among them, protection against radiation from natural as well as human-made sources stands out as one of the most desirable features. Thus, sun protection claims are not exclusive to sunscreen lotions but are also demanded in makeup compositions such as lipsticks and foundations. Furthermore, sun-blocking properties are generally achieved by the inclusion of ultraviolet (UV) filters in cosmetic formulas [2, 3].

UV filters are incorporated into cosmetic formulas to build up a protective screen that blocks most photons from penetrating the skin. UV radiation negatively affects skin cells by promoting DNA damage mediated by reactive oxygen species (ROS), by chemical modification that includes pyrimidine dimer formation, or by inducing replication errors, which can lead to erythema or ultimately increase the risk of melanoma [4, 5]. As UV radiation comprises three main subclasses of radiation based on their wavelength (UVA for the lowest energy photons ranging from 320 to 420 nm; UVB, in the middle range from 280 to 320 nm; and UVC for high energy photos with wavelengths from 100 to 280 nm), different chemical compounds are used to cover those ranges as few materials are completely effective to cover the entire UV region. Organic molecules such as benzophenone and avobenzone confer protection from UVA while others such as PABA derivatives and octocrylene protect from UVB radiation. UVC is generally not considered when designing cosmetic formulas because most of this type of radiation is lost at the upper layers of the atmosphere, where it is involved in ozone synthesis. Inorganic compounds are also used as UV filters as some of them possess broad spectrum blocking properties, which include kaolin, talc, calcium carbonate ($CaCO_3$), zinc oxide (ZnO), and titanium dioxide (TiO_2) [6, 7].

Sustainability initiatives focused on preserving the environment have spotlighted the raw materials used to produce many of the goods we consume regularly. The cosmetic industry is no stranger to such scrutiny, and specifically, recent studies have dealt with the potential threats organic UV filters pose to water sources, coral reefs, and the food chain [8]. Future directions point to an increase in the relevance of inorganic filters as an alternative to organic compounds to act as physical blockers of UV radiation in cosmetic formulas, as they are currently considered less harmful to the environment, and they pose, in general, a low risk to human health [9]. The US Food and Drug Administration (FDA) approves the use of TiO_2 and ZnO as photoprotective agents up to a limit of 25% w/w. They are commonly combined in formulas to give wide-spectrum protection, but a major drawback of this is the undesired white residue they leave when applied on the skin. This can be solved by working with engineered nano-sized ingredients for which this effect is absent. Nevertheless, some environmental and human health concerns are associated with nano-sized TiO_2 and ZnO, but more data is needed to fully assess their effects [10].

Quantification of TiO_2 is important not only to ensure regulatory compliance but also to control the batch-to-batch quality of cosmetic products containing it to claim UV-blocking properties or even use it as a pigment. Quantification is particularly challenging because of the complexity of cosmetic formulas [11], and, in contrast with studies concerned with the analysis of organic filters, not many literature works are devoted to the analysis of TiO_2 in such matrices [12]. TiO_2 quantification can be achieved via redox titration [13], portable X-ray fluorescence (pXRF) [14, 15], Raman spectroscopy [16], laser-induced breakdown spectroscopy [17], ICP/AAS approaches [18–23] and nanomaterial-based approaches [24]. Remarkably, most of these methods were developed for emulsion matrices only, which leaves a gap for the analysis of makeup matrices, which might involve additional considerations such as the presence of inorganic compounds commonly used as pigments.

When analyzed by methods such as AAS or ICP, samples that contain TiO_2 are processed with aggressive treatments that normally involve a digestion step with hydrofluoric acid (HF) or mixtures of HF with other reactive species such as concentrated nitric acid (HNO_3) and hydrogen peroxide H_2O_2 [25]. Other protocols use sulfuric acid (H_2SO_4) [26], which provides high recovery yields of TiO_2, but may lead to the formation of sulfur oxide species (S—O) with a similar m/z ratio as the main Ti isotope, which is inconvenient in some applications [27]. Different approaches include the formation of emulsion slurries before injection in the ICP [19, 20, 22]. For this study, we wanted to implement a borate fusion method, which has been used in treating geological and environmental samples, where it has shown better recovery compared to HF digestion methods [28–30] but, to the best of our knowledge, has not been implemented and validated for the analysis of cosmetic samples. In this approach, the mixture is pre-oxidized to remove organic interferences and then dissolved at the melting temperature of the borate salt in a platinum crucible (flux). Different borate salts are available to dissolve inorganic oxides depending on their acidity, and mixtures are normally used to work with samples that contain more than one type of oxide [31]. This is especially important for samples such as makeup, where inorganic oxides such as Fe_2O_3 and ZnO are used as pigments or UV filters.

For the quality assessment of cosmetic products, specifically in compliance with current regulations that set a maximum limit for TiO_2 of 25% w/w, we are mainly concerned with the development of methods that facilitate the routine analysis of products that have been designed to contain important amounts of inorganic filters. This is in clear contrast with methods aimed at the analysis of traces, where LOD and LOQ parameters are expected to be as low as the methods permit it. In this study, we sought to develop and validate a cost-effective method for TiO_2 extraction from emulsion, foundation, and lipstick matrices, and its quantification using FAAS, which is an affordable and widely implemented analytical technique for most quality control laboratories in the cosmetic industry. We assessed the specificity, linearity, precision, accuracy, and robustness of the method following ICH and USP guidelines [32, 33].

2. Materials and methods

2.1 Reagents and materials

Titanium 1000 mg/mL standard solution (Certipur®), reagent grade potassium chloride, reagent grade anhydrous aluminum chloride, 66% di-lithium tetraborate/34% lithium metaborate (SPECTROMELT ® A 12), and analytical grade 65% HNO_3 (EMSURE ®) were purchased from Merck (Darmstadt, Germany). TiO_2 raw material (UV-Balance Powder 100™) was obtained from KOBO Industries (New Jersey, USA). Lipsticks, liquid foundations, and cream emulsions were obtained from Belcorp (Lima, Perú). Deionized water (resistivity ≥18 MΩ*cm) was prepared with a Barnstead™ Easy Pure™ II water purification system.

2.2 FAAS conditions

Measurements were performed using a Thermo Scientific™ iCE™ 3000 atomic absorption spectrometer with an air-acetylene flame. Acetylene flux was 4.6 L min^{-1}. Ti absorbance was measured using a Hollow Cathode Lamp at a wavelength of 365.4 nm, with a bandpass of 0.5 nm. A deuterium lamp was used for background correction.

2.3 Sample preparation

Cosmetic matrix (lipstick, foundation, and emulsion) effects were studied through the analysis of placebos (no added TiO_2) specially prepared for this study. Calibration curves were completed for each matrix by combining placebo stock solutions and an appropriate amount of titanium standard solution. A defined amount of placebo (2000 mg of lipstick, 1400 mg of emulsion, 1200 mg of foundation) was weighted in a platinum crucible and mixed with 0.6 g of Spectromelt ® A 12. The crucibles were heated at 800°C for 1 h in a muffle furnace. Later, they were further heated at 1100°C for 15 min. After slow cooling to room temperature, the crucibles were transferred to 100-mL beakers, and a volume of 70 mL of 1 M HNO_3 was added. The solution was stirred with a magnetic bar at 80°C until complete dissolution of the solid residue. The cool solution was transferred to a 100-mL volumetric flask and completed to volume with 1 M HNO_3 to obtain the placebo stock solutions. Calibration standards for each matrix were then prepared by mixing 5.0 ml of placebo stock, 2.5 mL of 10% KCl, and 1.0 mL of 5% $AlCl_3$ with the appropriate quantity of titanium standard solution. The mixture was diluted to volume with 0.6% Spectromelt ® A 12 in 1 M HNO_3. For system and raw material calibration curves, deionized water was used instead of placebo stocks. Cosmetic and raw material samples were prepared by placing an accurately weighed amount, between 10 and 500 mg, in a platinum crucible. The contents were extracted following the protocol already described. The resultant mixture was filtered with a 0.45 μm nylon syringe filter. Next, 1 ml of the filtered solution was transferred to a 100-mL volumetric flask, mixed with 5 mL 10% KCl, 2 mL 5% $AlCl_3$, and diluted to volume with 1 M HNO_3 for FAAS analysis.

2.4 Method validation

The method was validated following ICH and USP guidelines [32, 33]. Specificity, linearity, precision, accuracy, and robustness assays were conducted as described below.

2.4.1 Specificity

Placebo blanks were prepared by following the entire extraction method from placebos to yield placebo stocks without further addition of Ti standard. Each blank was prepared in duplicate. To identify matrix interferences and assess the ability of the method to quantify Ti in their presence, the method specificity was determined by comparing the signal of placebo blanks with the background and the Ti signal.

2.4.2 Linearity and sensitivity

System linearity was evaluated by preparing three calibration plots of absorbance versus standard concentration (20, 35, 50, 65, and 80 ppm). The same procedure was repeated for each cosmetic matrix using placebo standards. Data regression and correlation significance were estimated using analysis of variance (ANOVA) and a Student's *t*-test, respectively. To evaluate sensitivity, the limit of detection (LOD) and limit of quantification (LOQ) were calculated using data from the regression analysis. LOD was calculated using the formula $3 * (\sigma/S)$, where σ is the intercept of the regression equation and S is its slope. Similarly, LOQ was computed using $10 * (\sigma/S)$. The LOQ was validated by measuring six replicates of stock dilutions or placebo stock dilutions to the calculated concentration and assessing their accuracy and precision. %RSD limit was set at 5.0%.

2.4.3 Precision

The method precision was determined considering repeatability, intermediate precision, and reproducibility. Repeatability was assessed by measuring six replicates in low, middle, and high concentration levels (each level in duplicate) for the system and each cosmetic matrix. System levels were prepared from the stock solution by dilution to a final concentration of 20, 50, and 80 ppm. Cosmetic matrix levels were prepared from their corresponding placebo stock by adding an appropriate amount of titanium standard and diluting to a final concentration of 3.2, 4.0, and 4.8% w/w for lipstick, 4.8, 6.0 and 7.2% w/w for emulsion and 5.6, 7.0 and 8.4% w/w for foundation. For raw material repeatability, only 1 level, the nominal concentration declared in its certificate of analysis, was considered. Six replicates of this level were prepared, as previously described, and measured. %RSD maximum limit was established at 5.0% for each concentration level. Intermediate precision was estimated at the nominal concentration from three cosmetic matrix samples and raw material samples. The sample duplicates were measured in two different days and by two different analysts. %RSD maximum intra-day and inter-day limits were set at 5.0%, whereas the global %RSD maximum limit was set at 8.0%, following USP validation criteria [33]. Reproducibility was evaluated by measuring 3 sample preparations of each cosmetic matrix and raw material in two different laboratories.

2.4.4 Accuracy

Samples of each cosmetic matrix were prepared by supplementing placebo stocks with titanium standard in three concentration levels (low, middle, high) and extracted following the procedure previously described. Raw material samples were analyzed assuming the concentration of TiO_2 reported in the certificate of analysis as the nominal concentration. Three samples for each concentration level were measured in duplicate. The accuracy was evaluated based on the recovery rate for each matrix.

2.4.5 Robustness

Robustness was assessed using a five-variable Youden-Steiner model [34]. Cosmetic matrix and raw material samples were studied under method standard conditions and alternative conditions in eight experiments, as described in **Table 1**. Samples were measured in duplicate for each set of conditions. Critical variables were defined as those with absolute differences greater than $\sigma * \sqrt{2}$, where σ represents the standard deviation calculated from the repeatability assays.

3. Results and discussion

3.1 Method development

Cosmetic products such as lipsticks, foundations, and emulsions contain different ingredients of both organic and inorganic nature [35]. For TiO_2 extraction from such matrices, the use of Spectromelt ® A 12, a 66% di-lithium tetraborate (LiT)/34%

Parameter	Experiment number[a]							
	1	2	3	4	5	6	7	8
Spectromelt ® A 12 (g)	0.7	**0.5**	**0.5**	0.7	0.7	**0.5**	0.7	**0.5**
Muffle furnace initial temperature (°C)	820	820	**780**	**780**	820	**780**	**780**	820
Muffle furnace initial step (min)	**40**	**40**	**40**	80	80	80	**40**	80
Muffle furnace final step (min)	**10**	20	20	20	**10**	**10**	**10**	20
Final concentration (ppm)	**30**	**30**	70	70	70	**30**	70	**30**

[a]*Bolded values are the standard conditions of the method.*

Table 1.
Difference values for the variables studied under the Youden–Steiner model.

lithium metaborate (LiM) flux, common to geological and environmental sample treatment [36] but not so common outside this field [37], is proposed as an alternative to the use of hazardous acids such as HF, or mixtures of strong acids and oxidizing reagents under harsh conditions. The ratio LiM/LiT is chosen considering that basic oxides such as ZnO, present in cosmetic formulas to boost UV protection, are more soluble in LiT, whereas acidic oxides such as TiO_2 itself are soluble in LiM [31]. The first step of calcination at 800°C is performed in a muffle furnace to remove organic interferences. The next step is dissolving the sample in the melted flux, for which further heating is carried out at 1100°C. Combining these two steps yields a solid residue easily attacked by diluted acids such as 1 M HNO_3. Flux mass is an important factor for this method as it impacts the extent of solubilization of the oxides. In general, sample/flux ratios varying from around 2–20 are used, and for this method, this ratio can vary in the range of 1–10. Higher ratios must rely on the assumption of high purity of the flux, as any impurities would be magnified in the sample analysis. Melting temperature and time are also important as temperatures exceeding 1100°C might yield loss of sample and flux due to volatilization or spreading of the mixture outside of the crucible. Finally, oxidation time is considered relevant as well because only oxides are soluble in the mixture. Thus, complete oxidation of other components must be assured [31]. Consequently, these factors were studied under an experimental design to test the robustness of the method, as described below. The method scope can easily be extended to raw material samples, specifically the UV-Balance Powder 100™, and similar preparations due to their relatively simple formulas compared to the more complex cosmetic matrices.

3.2 Method validation

3.2.1 Specificity

Background and placebo blank signals, plotted in absorbance units as a function of time, are presented in **Figure 1**. Signals from placebo blanks show no difference compared to the background signal. This result confirmed that the extraction procedure removes matrix interferences if existing and that the final concentration of excipients is low enough to have negligible effects on the stability of the signal, as can be seen from their progression in time. **Figure 1** also presents the Ti signal at the lowest concentration (20 ppm), showing that placebo signals do not interfere with Ti quantification.

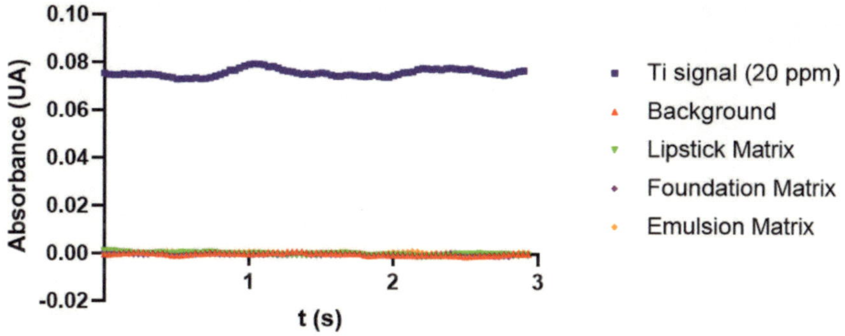

Figure 1.
Method specificity. Absorbance as a function of time for background, placebo blanks, and Ti (20 ppm) signals.

3.2.2 Linearity and sensitivity

The calibration plots obtained for raw material, lipstick, foundation, and emulsion matrices were adjusted using Least Squares Regression and judged regarding the value of the r^2 parameter, which was chosen as the acceptance criterion for linearity with a limit value of 0.99 and is presented in **Figure 2**. **Table 2** presents the values for r^2 and the other parameters of the model. According to these values, the data is well-fitted by a linear correlation in the range of concentrations used in this study. Furthermore, the statistical significance of these results, tested via ANOVA and Student's test, is demonstrated. Calculated LOD and LOQ values are presented in **Table 2.** LOQ precision is within limits defined following USP guidelines (%RSD $\leq$ 5.0%) [33], and, regarding the accuracy, the percentages of recovery at LOQ are not statistically different from 100%, as shown by using a Student's t-test. In general, the quality of these parameters illustrates the suitability of FAAS for the quantification of TiO_2 in cosmetic products.

3.2.3 Precision

Table 3 summarizes the total variation in terms of %RSD observed in this study for repeatability, intermediate precision, and reproducibility. %RSD values below 4.0% are observed across all concentration levels for all cosmetic matrices and below 3.0% for the raw material, showing that the method exhibits good overall repeatability. Regarding intermediate precision, an %RSD below 3.0% and near 1.0% for the foundation matrix indicates that the method is not significantly affected by slight changes in environmental conditions or different analysts. Regarding reproducibility, %RSD values are well below the USP recommended limit of 8.0%, which shows that the protocol was smoothly transferred to a different laboratory without significant loss in precision. %RSD values are comparable and, in some cases, lower than the results of other techniques [21], which is noteworthy as those methods are mainly designed for sunscreen preparation, whereas this method covers different categories of cosmetic products.

3.2.4 Accuracy

As an acceptance criterium for the accuracy of the method, a recovery rate not statistically different from 100% and contained in the interval 95–105% was defined as desirable. **Table 4** presents recovery rates for all cosmetic matrices and the raw

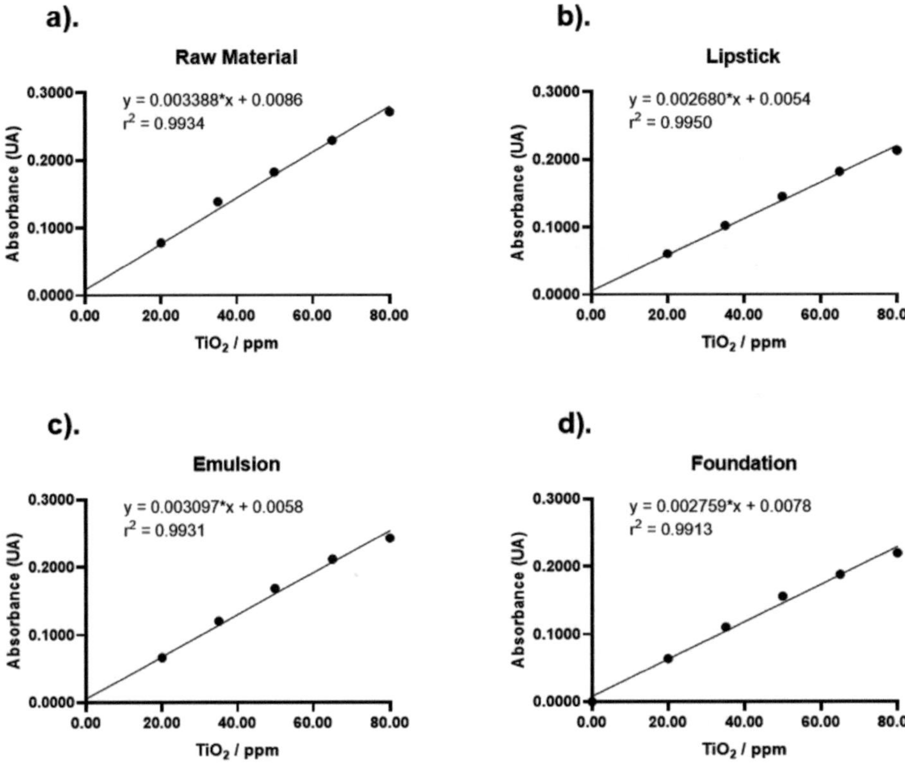

Figure 2.
Calibration plots for the raw material (a), lipstick (b), emulsion (c), and foundation (d) matrices.

		Raw material	Lipstick	Emulsion	Foundation
Linear range	Slope	0.003388	0.002680	0.003097	0.002759
	r^2	0.9934	0.9950	0.9931	0.9913
	t value[a]	71.8	82.1	69.9	62.1
	F value[b]/10^3	5.15	6.74	4.89	3.85
Sensitivity	LOD mg L^{-1}	2.2794	1.9935	2.3396	2.6353
	LOQ mg L^{-1}	6.9073	6.0409	7.0898	7.9859
	• %RSD	2.11	0.91	2.96	2.17
	• %Recovery	99.3	99.1	100.0	97.8
	• t value[a]	0.824	2.515	0.016	2.530

[a]*t critical value: 2.032 (linearity), 2.571 (LOQ).*
[b]*F critical value: 4.1366.*

Table 2.
Linearity and sensitivity parameters for all cosmetic matrices.

material. Data were evaluated by comparing nominal and found concentrations of TiO_2. The results show that the extraction protocol is optimal to achieve a variability below 3.0% for all matrices and concentration levels, except for the lowest level from foundation matrices where variability is below 5.0%. Global recovery rates are not significantly different from 100%, as demonstrated by Student's t-test t values.

		%RSD			
	Level	Raw material	Lipstick	Emulsion	Foundation
Repeatability	1		3.68	2.89	3.57
	2	2.21[a]	1.12	3.48	2.22
	3		1.37	3.52	3.21
Global intermediate precision		1.78	2.73	1.98	0.98
Global reproducibility		2.90	2.50	2.30	1.60

[a]*Raw material was evaluated at its nominal concentration, as declared by the certificate of analysis.*

Table 3.
Total variation (%RSD) for the precision levels considered in this study.

3.2.5 Robustness

Flux mass, muffle furnace initial temperature, duration of muffle furnace initial and final steps, and the final concentration of the sample for FAAS analysis were considered relevant factors for the study of the method robustness. Method-predetermined values for these factors were defined as standard conditions and deviations from them, as shown in **Table 1**, were defined as alternative conditions under a Youden-Steiner model. Critical factors are given by values that are higher than the critical value of each cosmetic matrix, calculated as described elsewhere [34]. As can be observed from the results presented in **Table 5**, final concentration is a critical factor for all cosmetic matrices and the muffle furnace initial temperature. These findings are a consequence of the complexity of cosmetic matrices regarding the wide chemical variety of their formulas. The result is influenced by the temperature at which oxidation is carried out because incomplete oxidation of the organic matter present in the formulas might interfere with the solubilization of the oxides in the flux later on the sample preparation protocol. At the same time, the effect of final concentration on the result can be explained because increasing organic/inorganic load without increasing the time of the oxidation step might also yield incomplete oxidation, which later interferes with the solubilization in the flux. The effects seen on cosmetic matrices can be contrasted to those observed for the raw material, where increasing the final concentration of the solution for FAAS analysis does not increase the organic matter content to be oxidized, and for which a slight decrease in the oxidation step temperature does not have a significant impact because the amount of organic matter to oxidize, compared to cosmetic matrices, is not significant. In addition to these factors, the mass of flux is critical only for foundation matrices processing. This is reasonable considering that apart from inorganic UV filters such as TiO_2 and ZnO, these compositions usually include some inorganic pigment load and, specifically for the matrices considered in this study, pigments based on Fe_2O_3, an oxide which would also be solubilized in the flux and thus would affect the working sample/flux ratio of the method.

4. Conclusions

We developed and validated a new method for the extraction and FAAS quantification of TiO_2 in cosmetic matrices such as emulsions, foundations, and lipsticks.

	Raw material		Lipstick			Emulsion			Foundation	
Added concentration (%p/p)	87.30[c]	3.200	4.000	4.800	4.800	6.000	7.210	5.600	7.010	8.380
Average found concentration[a] (%p/p, %RSD)	88.30, 0.93	3.219, 2.93	4.058, 0.95	4.815, 1.30	4.855, 1.84	6.094, 1.04	7.231, 2.00	5.584, 4.10	6.832, 2.23	8.613, 1.06
Recovery rate (m ean ± SD)	101.15 ± 1.03	100.59 ± 3.28	101.45 ± 1.05	100.22 ± 1.42	101.12 ± 1.98	101.52 ± 1.07	100.35 ± 2.20	99.76 ± 4.51	97.48 ± 2.41	102.72 ± 1.20
Global recovery rate[b] (mean ± SD)	101.15 ± 1.03		100.76 ± 1.94			101.00 ± 1.65			99.99 ± 3.48	

[a]*Sample size n = 6.*
[b]*t values: 1.961 (raw material), 1.175 (lipstick), 1.824 (emulsion), and 0.013 (foundation). t critical value: 4.303 (raw material), 2.306 (lipstick, emulsion, and foundation).*
[c]*Purity (nominal concentration) declared in the certificate of analysis.*

Table 4.
Recovery rates for all the cosmetic matrices studied in this work.

Parameter	Difference[a]			
	Raw material	Lipstick	Foundation	Emulsion
Spectromelt ® A 12 (g)	1.761	0.015	**0.382**	0.180
Muffle furnace initial temperature (°C)	0.357	**0.120**	**0.234**	**0.410**
Muffle furnace initial step (min)	1.012	0.054	0.027	0.100
Muffle furnace final step (min)	1.116	0.063	0.068	0.090
Final concentration (ppm)	0.605	**0.134**	**0.377**	**0.540**
Critical value	2.725	0.064	0.214	0.300

[a]*Bolded values correspond to critical factors.*

Table 5.
Youden-Steiner design for relevant factors of the method.

The method was validated according to ICH and USP guidelines and showed to be specific, accurate, precise, and robust. Borate fusion was successfully implemented for the extraction of TiO_2 with no use of HF or harsh conditions for digestion and oxidation of the organic content in the sample, which translates into a safer method. We showed the method can also be extended to the analysis of raw materials and is easily transferred to other laboratories, which is ideal in the context of quality assurance. A method is thus a tool to guarantee the safety of the products according to FDA and European Commission regulations and to ensure batch-to-batch quality.

Acknowledgements

The authors would like to acknowledge the Research, Development & Innovation Department of Belcorp for the financial support and research opportunities to create methodologies that ensure the quality and safety of products to fulfill consumer expectations.

Conflict of interest

The authors declare that there is no conflict of interest regarding the publication of this paper.

Author details

Cristian Rosales and Johnbrynner García*
Belcorp, Research, Development and Innovation, Tocancipá, Colombia

*Address all correspondence to: johngarcia@belcorp.biz

References

[1] L'ORÉAL Corp. 2019 Annual Report. Paris: L'ORÉAL Corp; 2019. Available from: https://www.loreal-finance.com/system/files/2020-03/LOREAL_2019_Annual_Report_3.pdf

[2] Lionetti N. New age of sustainable sunscreens: Formulating for modern sun care trends. Cosmetic Toilet. 2019;**134**:57-63. Available from: https://www.cosmeticsandtoiletries.com/formulating/category/suncare/New-Age-of-Sustainable-Sunscreens-Formulating-for-Modern-Sun-Care-Trends-558839411.html

[3] Coronado M. Evolving Trends and Hottest Ingredients in Sun Protection. 2017. Available from: https://blog.euromonitor.com/evolving-trends-hottest-ingredients-sun-protection/ [Accessed: July 1, 2020]

[4] Wilson BD, Moon S, Armstrong F. Comprehensive review of ultraviolet radiation and the current status on sunscreens. The Journal of Clinical and Aesthetic Dermatology. 2012;**5**:18-23

[5] Mark Elwood J, Jopson J. Melanoma and sun exposure: An overview of published studies. International Journal of Cancer. 1997;**73**:198-203. DOI: 10.1002/(SICI)1097-0215(19971009)73:2<198::AID-IJC6>3.0.CO;2-R

[6] Latha MS, Martis J, Shobha V, Shinde RS, Bangera S, Krishnankutty B, et al. Sunscreening agents: A review. The Journal of Clinical and Aesthetic Dermatology. 2013;**6**:16-26

[7] Ngoc LTN, Van Tran V, Moon J-Y, Chae M, Park D, Lee Y-C. Recent trends of sunscreen cosmetic: An update review. Cosmetics. 2019;**6**:64. DOI: 10.3390/cosmetics6040064

[8] Schneider SL, Lim HW. Review of environmental effects of oxybenzone and other sunscreen active ingredients. Journal of the American Academy of Dermatology. 2019;**80**:266-271. DOI: 10.1016/j.jaad.2018.06.033

[9] Schneider SL, Lim HW. A review of inorganic UV filters zinc oxide and titanium dioxide. Photodermatology, Photoimmunology & Photomedicine. 2019;**35**:442-446. DOI: 10.1111/phpp.12439

[10] Hanigan D, Truong L, Schoepf J, Nosaka T, Mulchandani A, Tanguay RL, et al. Trade-offs in ecosystem impacts from nanomaterial versus organic chemical ultraviolet filters in sunscreens. Water Research. 2018;**139**:281-290. DOI: 10.1016/j.watres.2018.03.062

[11] Iwata H, Shimada K. Practice of designing cosmetic formulations. In: Iwata H, Shimada K, editors. Formulas, Ingredients Prod. Cosmet. 1st ed. Japan: Springer; 2013. pp. 113-217

[12] Chisvert A, Salvador A. Ultraviolet filters in cosmetics: Regulatory aspects and analytical methods. In: Chisvert A, Salvador A, editors. Anal. Cosmet. Prod. 2nd ed. Amsterdam: Elsevier; 2018. pp. 85-106

[13] So Kim Y, Kim B-M, Park S-C, Jeong H-J, Seop Chang I. A novel volumetric method for quantitation of titanium dioxide in cosmetics. International Journal of Cosmetic Science. 2007;**29**:139-140. DOI: 10.1111/j.1467-2494.2007.00369_2.x

[14] Melquiades FL, Ferreira DD, Appoloni CR, Lopes F, Lonni AG, Oliveira FM, et al. Titanium dioxide determination in sunscreen by

energy dispersive X-ray fluorescence methodology. Analytica Chimica Acta. 2008;**613**:135-143. DOI: 10.1016/j.aca.2008.02.058

[15] Bairi VG, Lim J-H, Quevedo IR, Mudalige TK, Linder SW. Portable X-ray fluorescence spectroscopy as a rapid screening technique for analysis of TiO_2 and ZnO in sunscreens. Spectrochimica Acta Part B Atomic Spectroscopy. 2016;**116**:21-27. DOI: 10.1016/j.sab.2015.11.008

[16] Oh C, Yoon S, Kim E, Han J, Chung H, Jeong H-J. Non-destructive determination of TiO_2 concentration in cream formulation using Raman spectroscopy. Journal of Pharmaceutical and Biomedical Analysis. 2010;**53**:762-766. DOI: 10.1016/j.jpba.2010.05.008

[17] Menneveux J, Wang F, Lu S, Bai X, Motto-Ros V, Gilon N, et al. Direct determination of Ti content in sunscreens with laser-induced breakdown spectroscopy: Line selection method for high TiO_2 nanoparticle concentration. Spectrochimica Acta Part B Atomic Spectroscopy. 2015;**109**:9-15. DOI: 10.1016/j.sab.2015.04.010

[18] Mason JT. Quantitative determination of titanium in a commercial sunscreen formulation by atomic absorption spectrometry. Journal of Pharmaceutical Sciences. 1980;**69**:101-102. DOI: 10.1002/jps.2600690131

[19] Salvador A, Pascual-Martí MC, Adell JR, Requeni A, March JG. Analytical methodologies for atomic spectrometric determination of metallic oxides in UV sunscreen creams. Journal of Pharmaceutical and Biomedical Analysis. 2000;**22**:301-306. DOI: 10.1016/S0731-7085(99)00286-1

[20] Zachariadis GA, Sahanidou E. Multi-element method for determination of trace elements in sunscreens by ICP-AES. Journal of Pharmaceutical and Biomedical Analysis. 2009;**50**:342-348. DOI: 10.1016/j.jpba.2009.05.003

[21] Bunhu T, Kindness A, Martincigh BS. Determination of titanium dioxide in commercial sunscreens by inductively coupled plasma-optical emission spectrometry. South African Journal of Chemistry. 2011;**64**:139-143

[22] Lu PJ, Huang SC, Chen YP, Chiueh LC, Shih DYC. Analysis of titanium dioxide and zinc oxide nanoparticles in cosmetics. Journal of Food and Drug Analysis. 2015;**23**:587-594. DOI: 10.1016/j.jfda.2015.02.009

[23] Velimirovic M, Wagner S, Monikh FA, Uusimäki T, Kaegi R, Hofmann T, et al. Accurate quantification of TiO_2 nanoparticles in commercial sunscreens using standard materials and orthogonal particle sizing methods for verification. Talanta. 2020;**215**:120921. DOI: 10.1016/j.talanta.2020.120921

[24] Benítez-Martínez S, López-Lorente ÁI, Valcárcel M. Determination of TiO_2 nanoparticles in sunscreen using N-doped graphene quantum dots as a fluorescent probe. Microchimica Acta. 2016;**183**:781-789. DOI: 10.1007/s00604-015-1696-0

[25] Packer AP, Larivière D, Li C, Chen M, Fawcett A, Nielsen K, et al. Validation of an inductively coupled plasma mass spectrometry (ICP-MS) method for the determination of cerium, strontium, and titanium in ceramic materials used in radiological dispersal devices (RDDs). Analytica Chimica Acta. 2007;**588**:166-172. DOI: 10.1016/j.aca.2007.02.024

[26] Watkins PS, Castellon BT, Tseng C, Wright MV, Matson CW, Cobb GP. Validation of a Sulfuric acid digestion method for inductively coupled plasma

mass spectrometry quantification of TiO_2 nanoparticles. Bulletin of Environmental Contamination and Toxicology. 2018;**100**:809-814. DOI: 10.1007/s00128-018-2336-2

[27] Weir A, Westerhoff P, Fabricius L, Hristovski K, Von Goetz N. Titanium dioxide nanoparticles in food and personal care products. Environmental Science & Technology. 2012;**46**:2242-2250. DOI: 10.1021/es204168d

[28] Delijska A, Blazheva T, Petkova L, Dimov L. Fusion with lithium borate as sample preparation for ICP and AAS analysis, Fresenius'. Zeitschrift Für Analytical Chemistry. 1988;**332**:362-365. DOI: 10.1007/BF00468816

[29] Roy P, Balaram V, Bhattacharaya A, Nasipuri P, Satyanarayanan M. Estimation of Ti, Zr, Nb, Hf, Ta, Th and U in beach placers and ferrodiorites by inductively coupled plasma-mass spectroscopy using lithium metaborate fusion digestion technique. Current Science. 2007;**93**:1122-1126. Available from: http://www.jstor.org/stable/24099172

[30] Taylor V, Toms A, Longerich H. Acid digestion of geological and environmental samples using open-vessel focused microwave digestion. Analytical and Bioanalytical Chemistry. 2002;**372**:360-365. DOI: 10.1007/s00216-001-1172-z

[31] Claisse F. Fusion and fluxes. Comprehensive Analytical Chemistry. 2003;**41**:301-311. DOI: 10.1016/S0166-526X(03)41009-X

[32] Borman P, Elder D. Q2(R1) validation of analytical procedures. In: Teasdale A, Elder D, Nims RW, editors. ICH Qual. Guidel. An Implement. Guid. 1st ed. Hoboken, New Jersey: Wiley; 2018. pp. 127-166

[33] <852> Atomic Absorption Spectroscopy, United States Pharmacop. USP 41 NF 36. (2018) 6645-6648

[34] Karageorgou E, Samanidou V. Youden test application in robustness assays during method validation. Journal of Chromatography. A. 2014;**1353**:131-139. DOI: 10.1016/j.chroma.2014.01.050

[35] Baki G, Alexander KS. Introduction to Cosmetic Formulation and Technology. 1st ed. New Jersey: Wiley; 2015

[36] Makombe M, van der Horst C, Silwana B, Iwuoha E, Somerset V. Optimisation of parameters for spectroscopic analysis of rare earth elements in sediment samples. Rare Earth Elements. 2017;**2017**:74-78. DOI: 10.5772/intechopen.68280

[37] Braysher E, Russell B, Woods S, García-Miranda M, Ivanov P, Bouchard B, et al. Complete dissolution of solid matrices using automated borate fusion in support of nuclear decommissioning and production of reference materials. Journal of Radioanalytical and Nuclear Chemistry. 2019;**321**:183-196. DOI: 10.1007/s10967-019-06572-z